www.piratemultiplication.com

100
pastimes to revise in 25 different ways the times tables and the divisions

"100 pastimes to revise in 25 different ways the times tables and the divisions in order to strengthen your attention, perception, working-memory and numerical reasoning"

These pastimes will require the parents/teachers collaboration to be explained according to the proposed instructions.

Authorship
©Pedro Santos Juanes Muñoz
Graphic design and book layout
Pedro Santos Juanes Muñoz
Drawings:
Pirate: Pere Amorós Morató
Title Page (cave): Designed by upklyak / Freepik
Title Page (forest): Designed by brgfx / Freepik

WEB:
www.piratemultiplication.com

The video game download code is on page 72

E-mail:
hello@piratemultiplication.com

edition 1, May 2021

These materials have been developed by Pedro Santos Juanes Muñoz (2020), and thereby accordingly registered. Therefore, their creation mustn't be attributed to anyone, except the author. No part of this book may be reproduced, translated, stored in a retrieval system, or transmitted, in any form or by any means, electronic, mechanical, photocopying, microfliming, recording, or otherwise, without written permission from the author
hello@piratemultiplication.com

INDEX

The story...pg 4
Instructions for the 25 games...pg 5

Activities	multiplications		division	
Routes	① pg 18	㉖ pg 31	㊿+1 pg 44	㊀ pg 57
Numbers Search Puzzle	② pg 18	㉗ pg 31	㊾+3 pg 44	㊆ pg 57
The Treasure	③ pg 19	㉘ pg 32	㊿+3 pg 45	㊇ pg 58
The ladder	④ pg 19	㉙ pg 32	54 pg 45	79 pg 58
The paintings	⑤ pg 20	㉚ pg 33	55 pg 46	80 pg 59
The magnifying glass	⑥ pg 20	31 pg 33	56 pg 46	81 pg 59
The barrels	⑦ pg 21	32 pg 34	57 pg 47	82 pg 60
The maze	⑧ pg 21	33 pg 34	58 pg 47	83 pg 60
"Boxing"	⑨ pg 22	34 pg 35	59 pg 48	84 pg 61
The swing	⑩ pg 22	35 pg 35	60 pg 48	85 pg 61
Little lanterns	⑪ pg 23	36 pg 36	61 pg 49	86 pg 62
Three in a raw	⑫ pg 23	37 pg 36	62 pg 49	87 pg 62
The bottle	⑬ pg 24	38 pg 37	63 pg 50	88 pg 63
The deposit box	⑭ pg 24	39 pg 37	64 pg 50	89 pg 63
The mirror	⑮ pg 25	40 pg 38	65 pg 51	90 pg 64
The keys	⑯ pg 25	41 pg 38	66 pg 51	91 pg 64
The exam	⑰ pg 26	42 pg 39	67 pg 52	92 pg 65
The patterns	⑱ pg 26	43 pg 39	68 pg 52	93 pg 65
The paths	⑲ pg 27	44 pg 40	69 pg 53	94 pg 66
The bomb	⑳ pg 27	45 pg 40	70 pg 53	95 pg 66
The domino game	21 pg 28	46 pg 41	71 pg 54	96 pg 67
The horse	22 pg 28	47 pg 41	72 pg 54	97 pg 67
Symmetries	23 pg 29	48 pg 42	73 pg 55	98 pg 68
Investigating	24 pg 29	49 pg 42	74 pg 55	99 pg 68
Diamonds	25 pg 30	50 pg 43	75 pg 56	100 pg 69

Answer Key...pg 72

The story...

Once upon a time, luckily, we discovered that a King messenger was hidden amongst the crew in one of the boats that we captured. He was secretly carrying an encrypted message, and as there was no way to understand it, we abducted an accountant being an expert in Mathematics to decipher the message. It said: "The King's boat will arrive at the usual place where he is owed all the taxes to be paid". And although it turned out to be evident that there was a lot of money in that place, we needed to be told where the exact meeting point would take place. Generally the King's private messengers are mutes and they don't know how to read or write in order to avoid muttering, but we made him try the "soap of the truth" and he confessed: he told us everything. The Red Raven went there and what we found was a huge amount of small ships so loaded with gold and jewellery that they hardly seemed to be able to stay afloat. All those ships were heading to the King's boat for the taxes to be paid. The spoils we got were so big that we had no other option but to abduct the accountant again for him to clarify to us the value of all of that. Moreover, according to the accountant, it would take us one year to make the necessary hole to hide the goods, and as it was not possible for us to be digging up for that period of time, we looked for a cave and we left the treasure there. We had to return the accountant to his house and I was left to guard the cave. Thank god the accountant gave me these pastimes. And although a lot of people don't believe it, we must know very well the multiplication tables to multiply treasures and divide the spoils.

Activity instructions:

ROUTES: 1, 26, 51, 76.

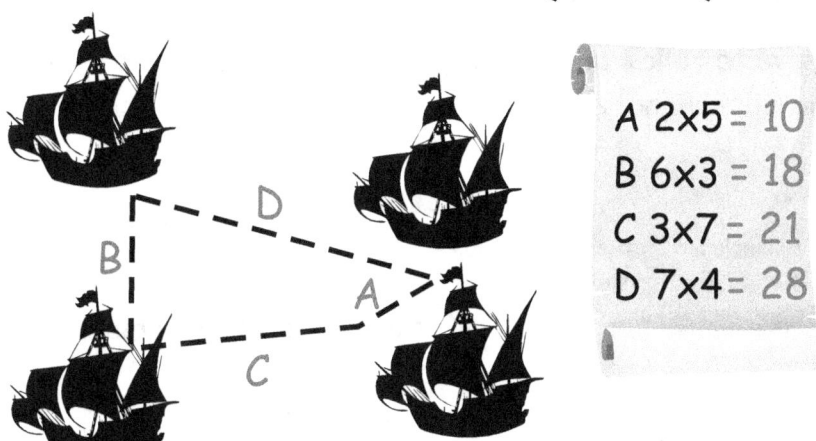

A 2×5 = 10
B 6×3 = 18
C 3×7 = 21
D 7×4 = 28

In the routes game, first, you have to solve the calculations (multiplications and divisions) that appear on the parchment paper and then write on every section of the shape that the ships define the letter that is suitable according to the length of the section. That is, the letter having the highest result will be placed on the longer section of the shape, whilst the letter having the shortest result will be placed on the shortest one.

SOUP: 2, 27, 52, 77.

In this activity you have to find five calculations (multiplications or divisions) that are all jumbled up and appear on the number search puzzle. This will be horizontally or vertically. We will have as an example one of the following options as you wish:
4×2 = 8 or 8: 2 = 4, 8×3 = 24 or 24: 3 = 8,
9×3 = 27 or 27: 3 = 9, 3×10 = 30 or 30: 10 = 3,
6×9 = 54 or 54: 9 = 6

4	2	8	7	35
2	3	8	3	24
9	6	4	54	5
3	10	30	9	4
27	8	3	6	10

Game instructions:
CHOOSING THE TREASURE: 3, 28, 53, 78.

In this activity you will have to follow one by one the instructions on the parchment paper, and you will eliminate the numbers that don't meet the conditions.

We will then eliminate number 25, 35, 45 ⇐ 1.- It is included on the 5 times tables ⇒ we will then eliminate number 18
2.- It is an Even number.
3.- It is > than 40 and < than 70 ⇒ We will eliminate number 20, being left number 50 as the chosen treasure

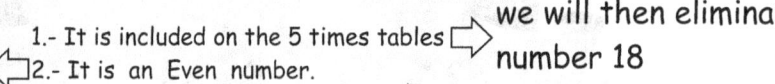

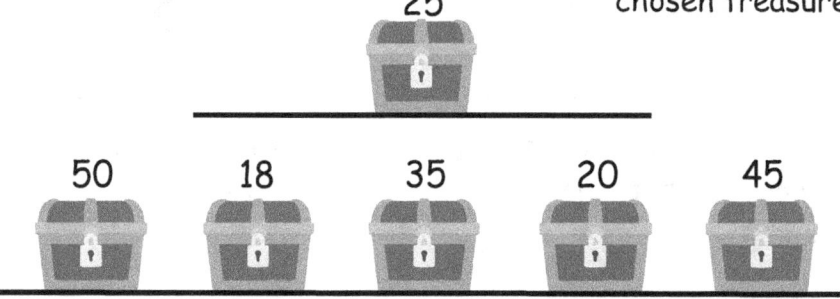

THE LADDER: 4, 29, 54, 79.

Firstly, solve the calculations and then place on every space of the ladder the appropriate multiplication. None of the numbers placed on the spaces of the ladder must be repeated on the above or below space. For example, if you have the following calculation: "7x3=21", the calculation that is placed on the above or below space mustn´t contain the numbers 1,2,3 or 7.

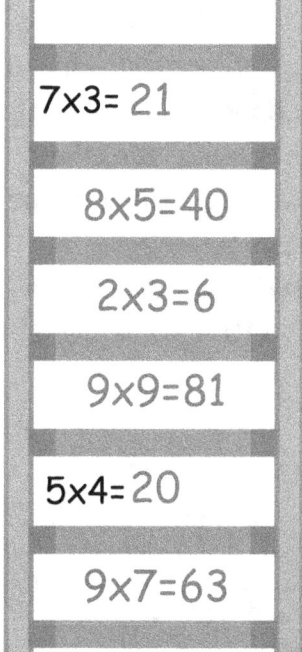

7x3= 21

8x5=40

2x3=6

9x9=81

5x4= 20

9x7=63

7 x 3 = 21
2 x 3 = 6
9 x 7 = 63
9 x 9 = 81
8 x 5 = 40
5 x 4 = 20

Game instructions:

THE PAINTINGS: 5, 30, 53, 78.

  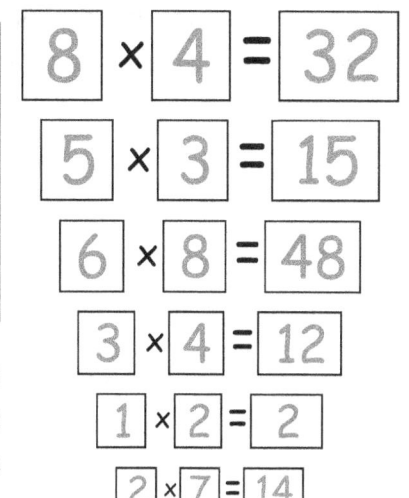

In this activity you have to match up the numbers according to their size and then solve the calculation (multiplication or division) as indicated and on the appropriate box. That is, the bigger numbers will be placed on the bigger boxes whilst the smaller ones will go on the smaller boxes.

THE MAGNIFYING GLASS: 6, 31, 54, 79.

Solve the calculation that you can find on the magnifying glass. For example, "2x7=14". Look at the arrow that is drawn on the magnifying glass. If it is pointed upward, you have to find the closest number to the above result. In this example, the number would be "15", so we will have to find on the 2 to 10 times tables a calculation whose result consists of that number (for example, "3x5=15"). However, if the arrow is pointed downward, you will have to find the closest number below the result, in this case the required number would be number 13. As there is no multiplication operation on the times tables (from 2 to 10) whose result consists of number 13, we keep finding downward and we move to number 12, which is a valid result: "2x6=12. In the division calculations you are not allowed to use the quotient time table and if you want to make it a bit more interesting, the use of the 2 to 10 times tables won't be allowed. You are not allowed to repeat the division to obtain the same number either.

In the divisions: 2̶ 1̶0̶

36:6= 6
      ↓
      5   20 : 4 = 5

You are not allowed to use number "6" times tables result

Game instructions:

THE BARRELS: 7, 32, 57, 82.

In this activity you have to solve the calculations that appear floating in the sea and write them on the different barrels by taking into consideration that the bigger the result is, the deeper the barrel is located.

2x5, 7x7, 4x8, 2x9, 5x6, 8x6

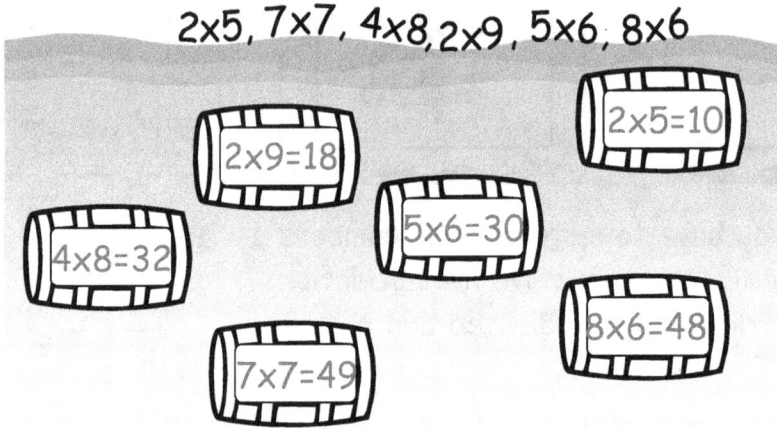

THE MAZE: 8, 33, 58, 83.

Find from one of the entrances the shortest path to get to the first pearl taking into account that the path goes left to right and that you are not allowed to pass the same path twice. Afterwards, find from that box the shortest path to get to the second pearl and do the same from the second to the third pearl. Once you are on the third pearl, find the closest exit (in the example: A or B). Count all the boxes you have gone through at this point from the beginning. Write it down on the final result box which appears underneath the maze (number 30 in the example). At this point find and write down on the rest of the boxes three multiplication or division calculations and their results according to the indicated symbol (X :) whose results by calculating them (adding them up, deducting them, multiplying them or dividing them) will allow you to obtain that result. If you want to do it more complicated, don't use the 1 times tables and just once the 10 times table.

Game instructions:

"BOXING": 9, 34, 59, 84.

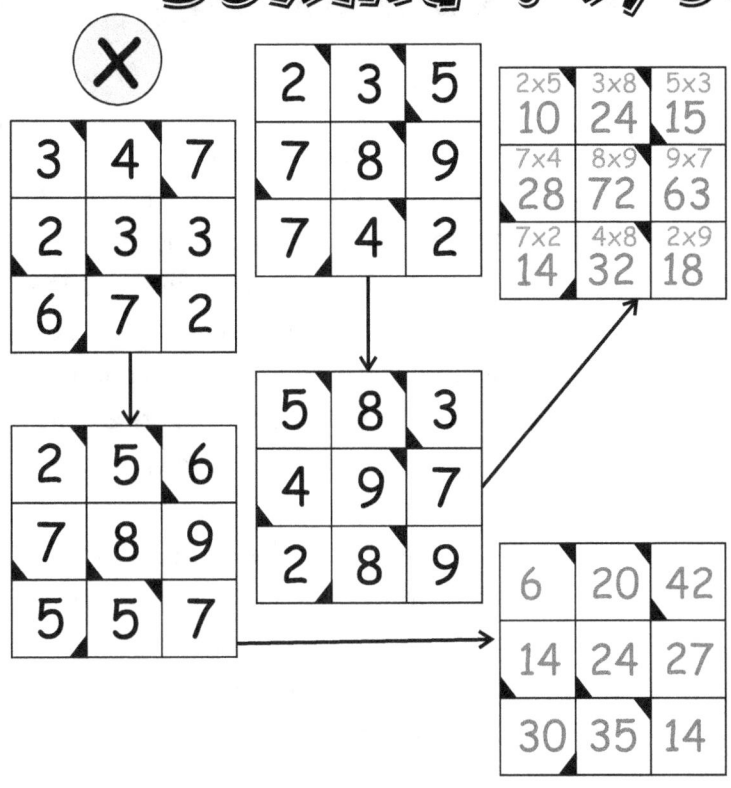

Find the grids which have the boxes decorated in the same way. Match them up with lines. Multiply or divide them according to the sign that you will see in the activity within a circle. Write down the result according to the order of its box. That is, the first number of the first box on the grid will be multiplied or divided by the number of the first box of the other grid, writing it down on the first box of the result grid, which is the one which has no numbers in it, and so on.

THE SWING: 10, 35, 60, 85.

1, 2, 3, 10, =

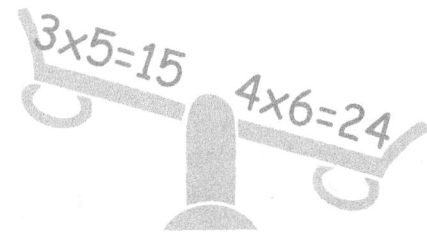

Find multiplication/division calculations which by placing them on the empty side of the swing allows to keep the logic of the picture. That is, if the right side falls below, that means that it weighs more than the left side and therefore the multiplication calculation on that right side will be bigger. In this activity, you are not allowed to use the 1, 2, 3 and 10 (or any other) times tables if the teacher thinks it is appropriate. You are not allowed to use the same times table on the same scale.

Game instructions:

LITTLE LANTERNS: 11, 36, 61, 86.

9x9 8x8 7x7
81   64   49

6x6   4x4
     5x5
36   25   16

Look at the multiplications that are above the little lanterns, they form a systematic sequence, that is, they follow a particular rule. Find out the rule that predicts which multiplication will be the following one. Once you work out the rule, apply it and write it down on the string that links the little lanterns and above them the multiplications that follow the sequence. Write down the answer to the multiplications in the centre of the little lantern.

THREE IN A ROW : 12, 37, 62, 87.

Solve the multiplications to find out if there is a three in a row, that is, the same result being linearly shown three times (horizontally, vertically, diagonally). Highlight the different threes in a row that you will find.

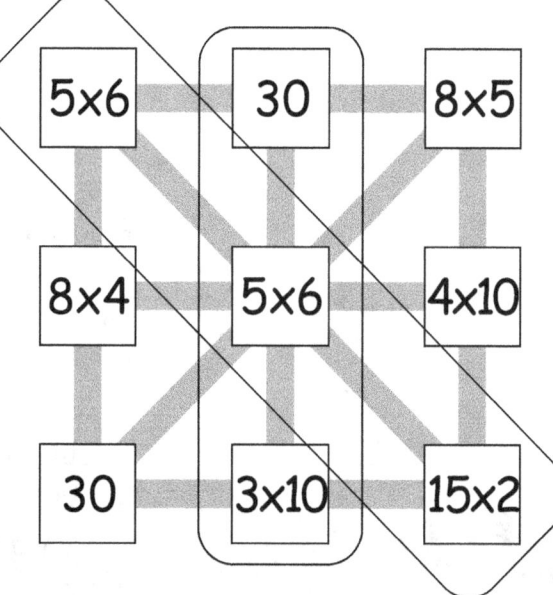

5x6	30	8x5
8x4	5x6	4x10
30	3x10	15x2

Game instructions:

THE BOTTLE: 13, 38, 63, 88.

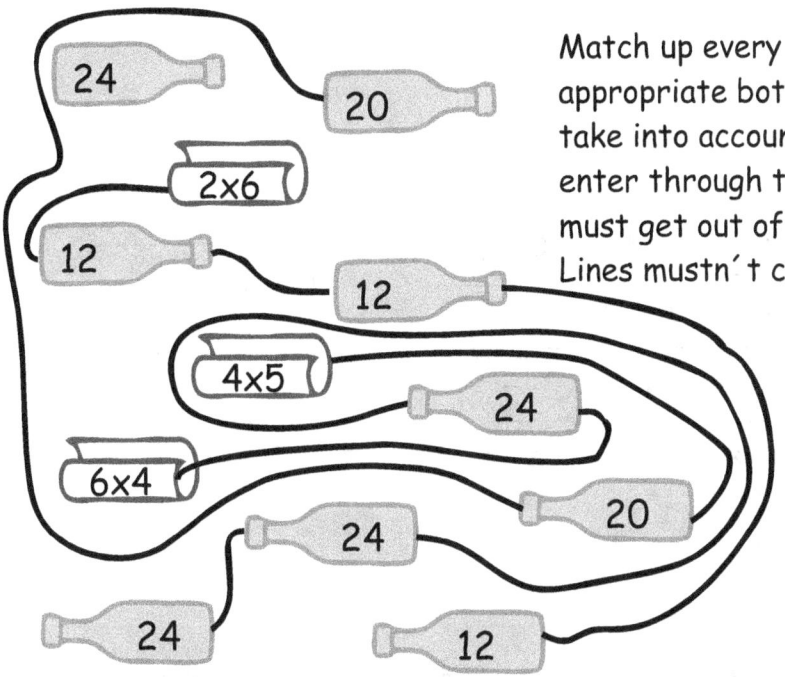

Match up every message to the appropriate bottle. However, you have to take into account that the lines must enter through the base of the bottle and must get out of the bottle neck finish. Lines mustn´t cross.

THE DEPOSIT BOX: 14, 39, 64, 89.

You have to use this code system to discover the numbers that will open the deposit box. You have to keep adding up or deducting as the pictures show. Solve the multiplications or divisions once you have discovered the numbers that are included in them.

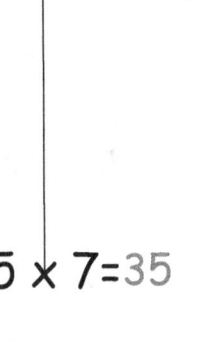
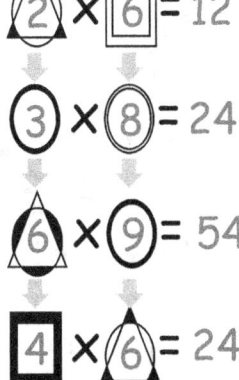

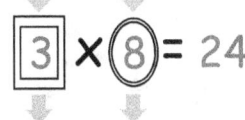

Game instructions:

THE MIRROR: 15, 40, 65, 90.

Solve the multiplications. Look at the results and find pairs whose result is the same if we swap the order of their two figures. Write them down on the mirrors. In the example we have number 12 and 21, number 42 and 24.

3 × 4 = 12   7 × 5 = 35
7 × 6 = 42   9 × 2 = 18
9 × 7 = 63   7 × 3 = 21
3 × 5 = 15   8 × 3 = 24

3x4=12
7x3=21

7x6=42
8x3=24

THE KEYS: 16, 41, 66, 91.

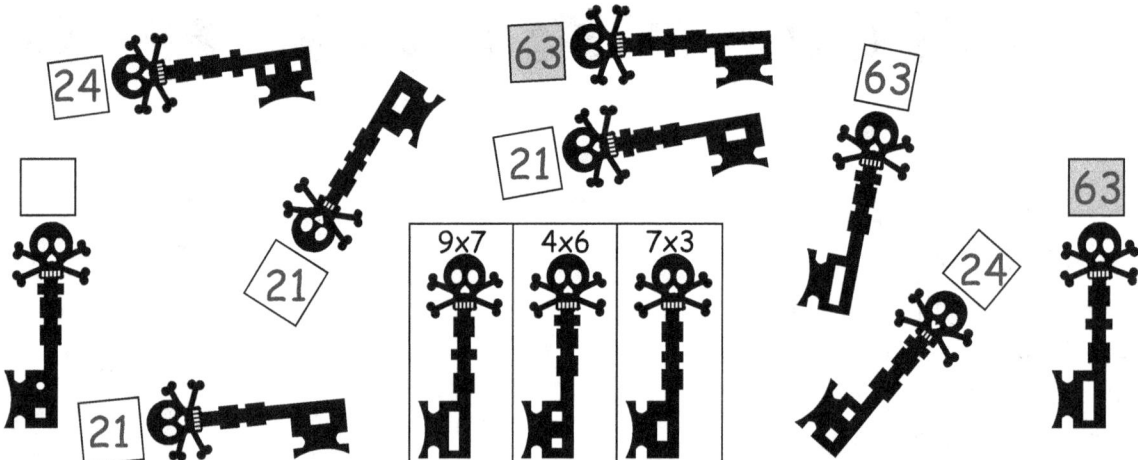

You will find a lot of types of keys, these keys have a little box on top of their skulls. You have to write inside this little box the result of the multiplication that the appropriate master key will indicate. Each masterkey will be identified by each own distinguishing bottom part which you can find in each big box in the center. If the key is identical in every other respect, you will have to colour in the little box or you will have to highlight it somehow. You can find one or more keys which are identical to the masterkeys; keys with the same bottom or top side or even keys which are totally different.

Game instructions:

EXAM: 17, 42, 67, 92.

Nobody has written down their names on the exam, so Pirate Cook will find the one which has the least amount of mistakes and will write down his name. Correct the exams and choose the one with the least number of mistakes. Highlight it with an arrow or emphasize the most suitable way for you.

PATTERNS: 18, 43.

You must create a pattern by following the sample. You must follow these steps to know which number you have to write down.

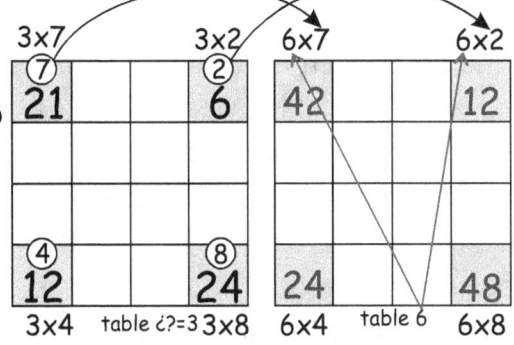

1st- From the numbers that appear on the pattern sample boxes, find out which times tables are being used. In the example the 3 times table is the right one.
2nd -Once you know which times table is the one which has been used, find out in each pattern box by which number had to be multiplied. Write down that number in the circles.
3rd-The number which has been used to multiply on each main pattern box is the one you have to use to multiply on the boxes of the following grids.

PATTERNS: 68, 93.

You will see the result which is needed on every grid. On each pattern box you will have to create a division whose result is that one, it is evident that you can't divide by 1 or 10. And if you want to make it more difficult, the divisor must be the same on each box/pattern.

	14:7	
18:9		
		12:6
	10:5	

Result 2

	21:7	
27:9		
		18:6
	15:5	

Result 3

-13-

Game instructions:

THE PATHS: 19, 44, 69, 94.

Solve the calculations writing down the result in the correct box. Follow the path to guess which is the right box. If you find it messy, it will be easier if you colour in the different paths.

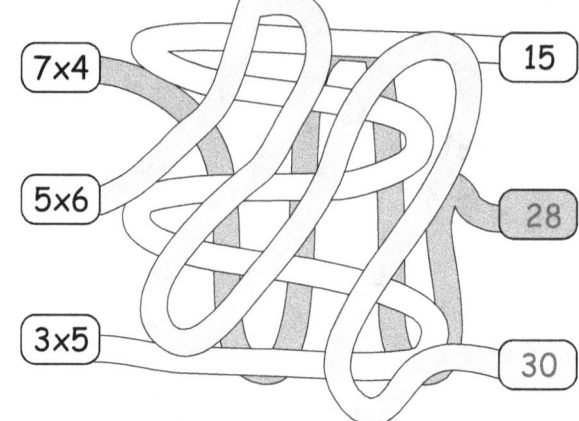

THE BOMB: 20, 45, 70, 95.

The stopwatch is running. You will have to quickly work on the numbers in the boxes surrounding the bomb in order to know how many minutes are left for the bomb to explode. You will have to find among the times tables (from the 2 to the 10) which numbers result in the ones which appear in the boxes. For example, if there is a number 10 in one of the boxes which surround the bomb, we know that by multiplying number 5,2 and 10 we get number 10 (2x5,10x1). These numbers will be written next to the different boxes. These numbers can't be bigger than 10, that is, you can't use 2x12 to get a number 24, either 3x15 to get a number 45. Once you have done this procedure with all the numbers, you will have to watch for the number which is the most repeated. In the example which is presented, the most repeated times table is the one corresponding to number 2, so we can affirm that the bomb will explode in 2 minutes. Write down inside the bomb this number, so it will get deactivated. When you finish all the bombs, stop the stopwatch. If you have been able to deactivate them before the shortest time set, you are a real bomb deactivator. If you finish before the average time set, you will be saved. However, if you haven't solved it before the longest time set, you will have to keep training because the explosion has been unavoidable. The ship crew relies on you, be quick but accurate working on the numbers!

Game instructions:

DOMINO GAME: 21, 46, 71, 96.

Complete the domino game. Take into consideration that the game requires matching dominoes. You will find a number in the middle of every tile. By dividing that number by one of the numbers which appear on the tile, you will get the missing number. In other words, the number which appears in the centre of the tile is obtained by multiplying both numbers which have been written on both domino ends.

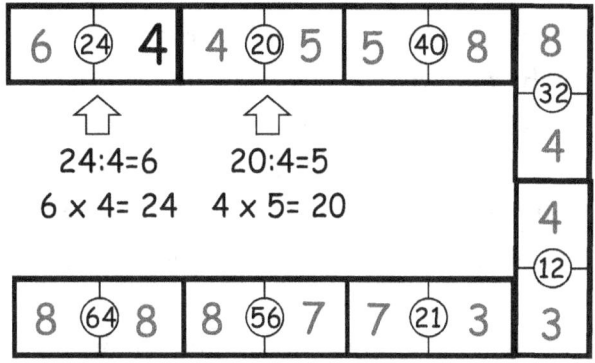

THE HORSE: 22, 47, 72, 97.

You have 10 seconds to guess which horse will win the race. You will have to focus on the moves that the black horse can make and the calculations that appear on the different boxes within the period of ten seconds. You will also have to focus on the boxes and calculations where the white horse can be moved. Make an estimate. Which horse will get more points? Cross with a "X" the horse that you think will get more points. The white horse has been crossed in the example. Solve the calculations and check whether your estimate is correct. Remember the moves that the horse makes on a chessboard.

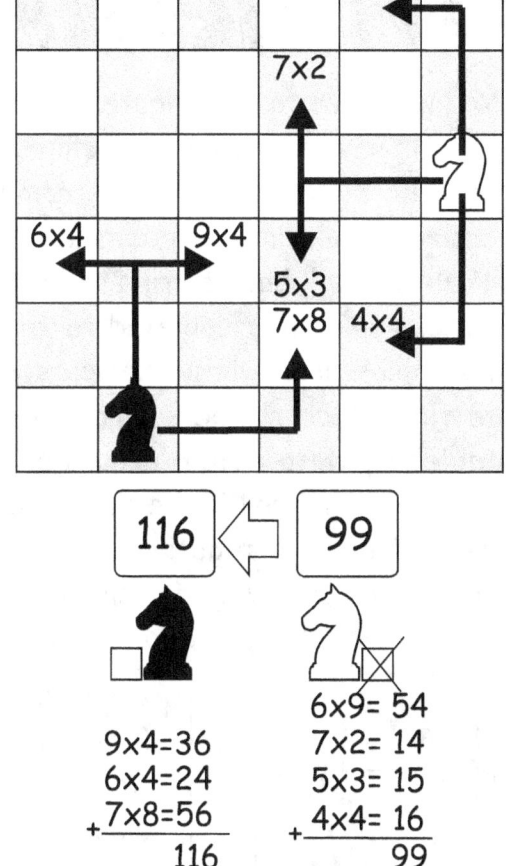

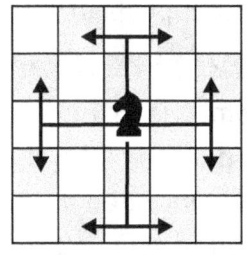

Game instructions:

SYMMETRIES: 23, 48, 73, 98.

In this activity you will find a drawing with two areas divided by a line consisting of a symmetry axis. On the upper part, you will find multiplications and a line which is linking them. On the bottom part, you will find just the dots. You will have to link the appropriate dots with a line by creating the symmetrical drawing which appears on the upper part (same distance to the axis). On the dots which you have linked by drawing the line, write down the result of the multiplications which are shown on the upper part.

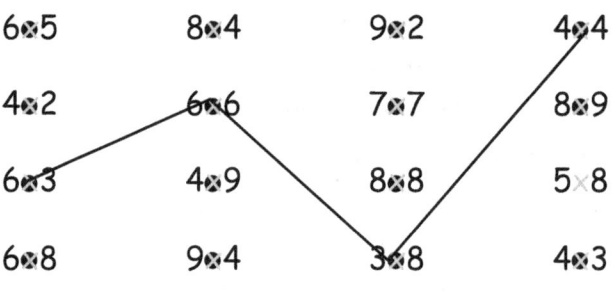

INVESTIGATING: 24, 49, 74, 99.

You will have to investigate which numbers you will have to place on the empty boxes to get the right result. First, complete the easier calculations. For example, (2x6=12). Then, you will have to find which different options could be completed for the rest of the calculations. In the example you will have to find which numbers (from the 4 times table) allow to obtain a number with the figure of number 1 in the first place (4x3 or 4x4). The investigation is about being consistent by trying different options and considering if that option allows you to appropriately finish the calculations. For example, 4x4=16. If this result is summed up to number 12 (resulting from the other calculation), you will get 28 which is not part of the 6 times table, so it will be discarded. You will need to try a different option (4x3) which allows you to square the results.

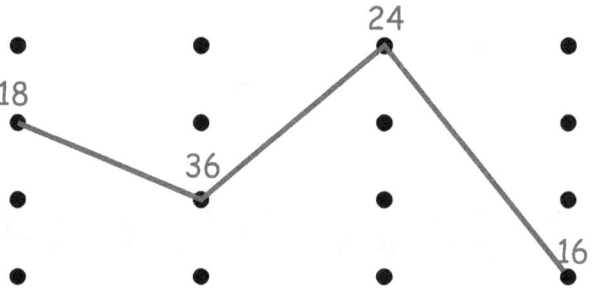

Game instructions:
DIAMONDS: 25 AND 50.

Guess how many gold coins you will get in the spoils. Find each type of diamond out of the bunch and write down how many there are of every kind. By knowing how much every diamond is worth in the market, find out how many coins you will get in the final spoils.

	◇	◇	◇	◇
value ○	9	4	6	3
there are	6	3	4	5
The value achieved with Each type of diamond will be:	6 x 9 54	3 x 4 12	4 x 6 24	5 x 3 15

54+12+24+15= 105

In this treasure I will be able to get ___105___ gold coins.

DIAMONDS: 75 AND 100.

In activities 75 and 100 you will see that in the table to complete there are two other rows of boxes. One of them is titled "If we are to distribute ..." where the pirates who have to distribute those gold coins are noted. The second row is titled "Each of us will touch ..." here you have to find out how many coins each pirate will touch.

The value achieved with each type of diamond will be:	6 x 9 54	3 x 4 12	4 x 6 24	5 x 3 15
If we are to distribute...	6	4	8	5
Each of us will touch ...	9	3	3	3
	↑ 54:6	↑ 12:4	↑ 24:8	↑ 15:5

1 Routes

A 3x8
B 4x9
C 7x3
D 7x5

2 Numbers Serarch Puzzle

12	2	6	5	32
2	3	8	3	24
5	3	4	12	5
3	10	34	9	4
15	7	2	5	10

 the treasure

1.- It is in the 7 times table.
2.- It is an even number
3.- It is > than 40 and < than 70
4.- It has a 5.

56

48   40   70   42   35

 The ladder

2x4=

5 x 8 =
8 x 6 =
9 x 3 =
2 x 4 =
3 x 7 =
3 x 5 =

5x8=

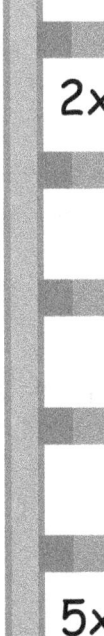

-19-

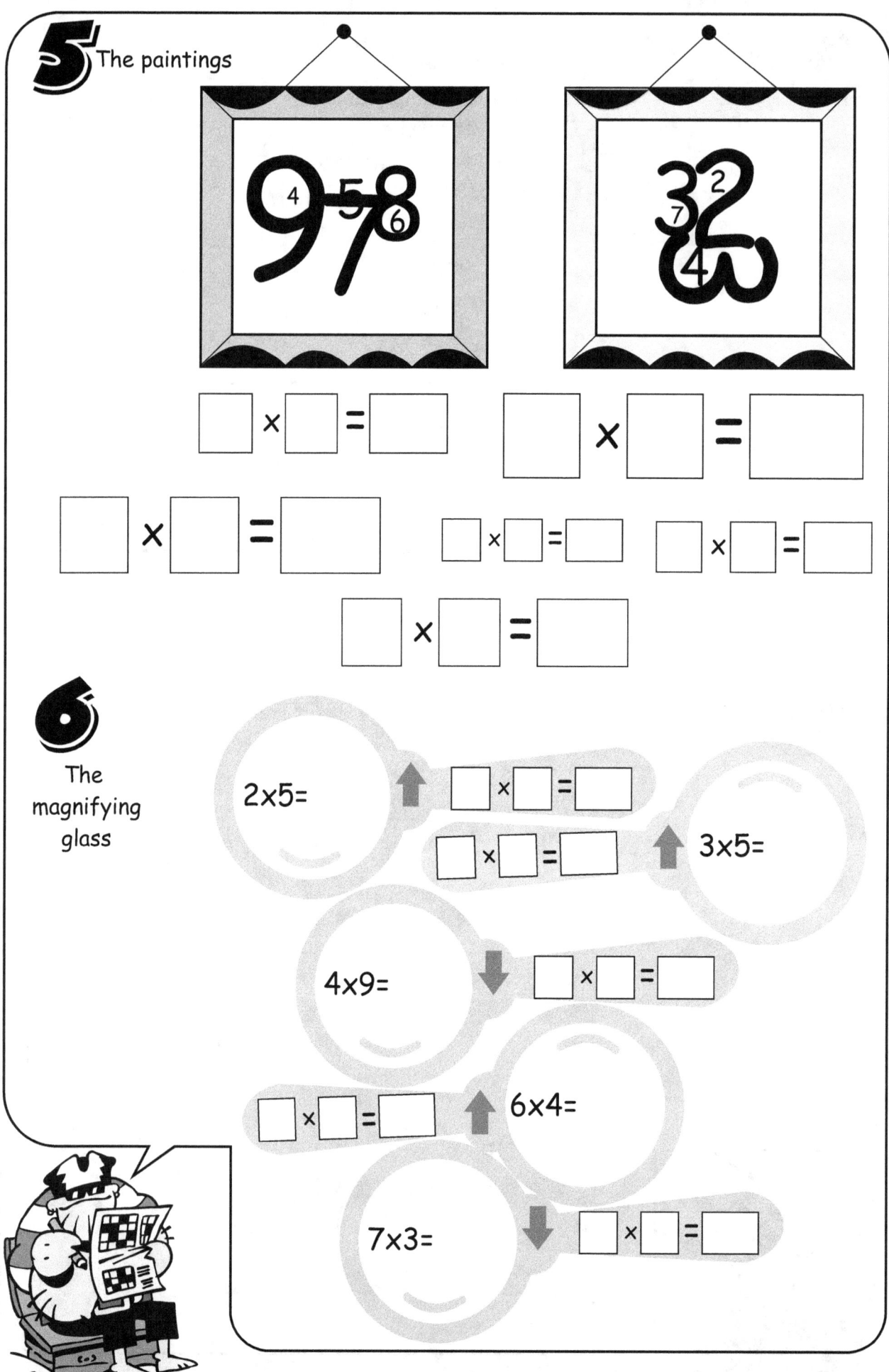

7 The barrels

8x4, 5x7, 4x9, 3x7, 6x4, 9x3

8 The maze

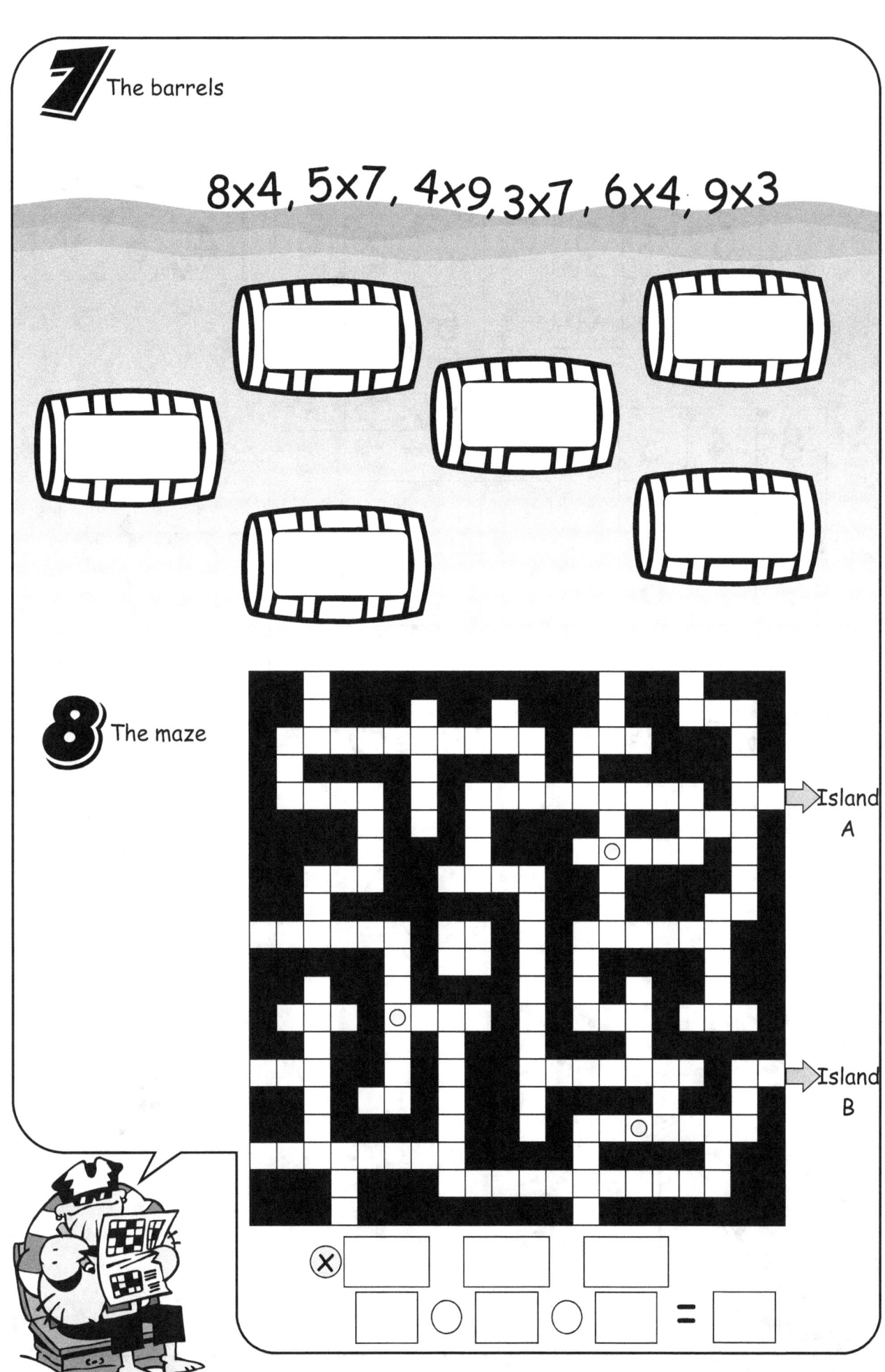

9 "Boxing" ×

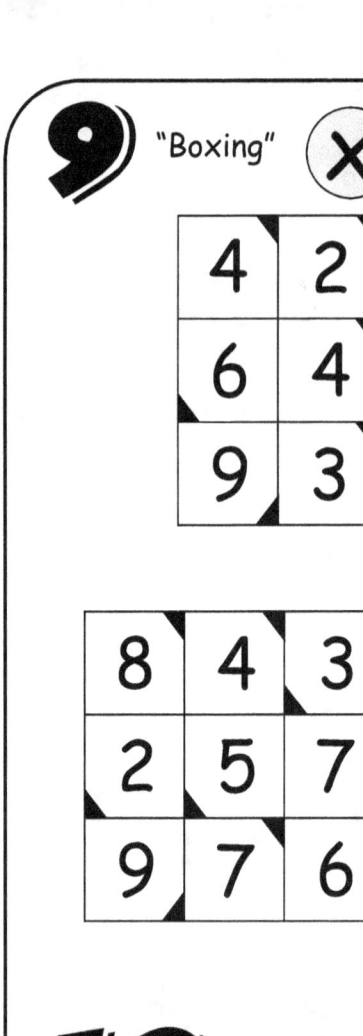
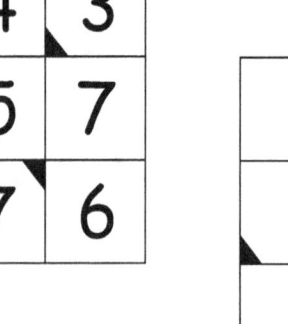
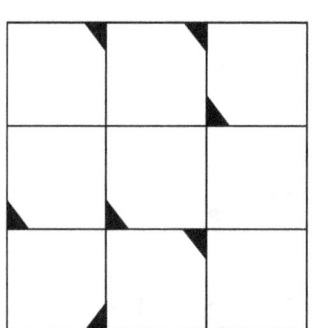
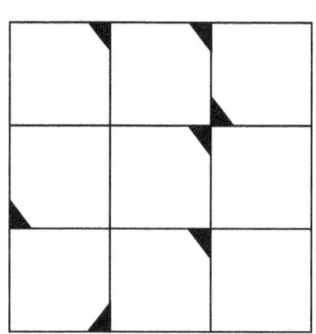

10 The swing

~~1, 2, 3, 10~~

27

56

27

42

49

11 Little lanterns

2x3 — 6
3x4
4x5
5x6

12 Three in a row

12	3x5	2x7
15	4x3	5x3
14	15	6x2

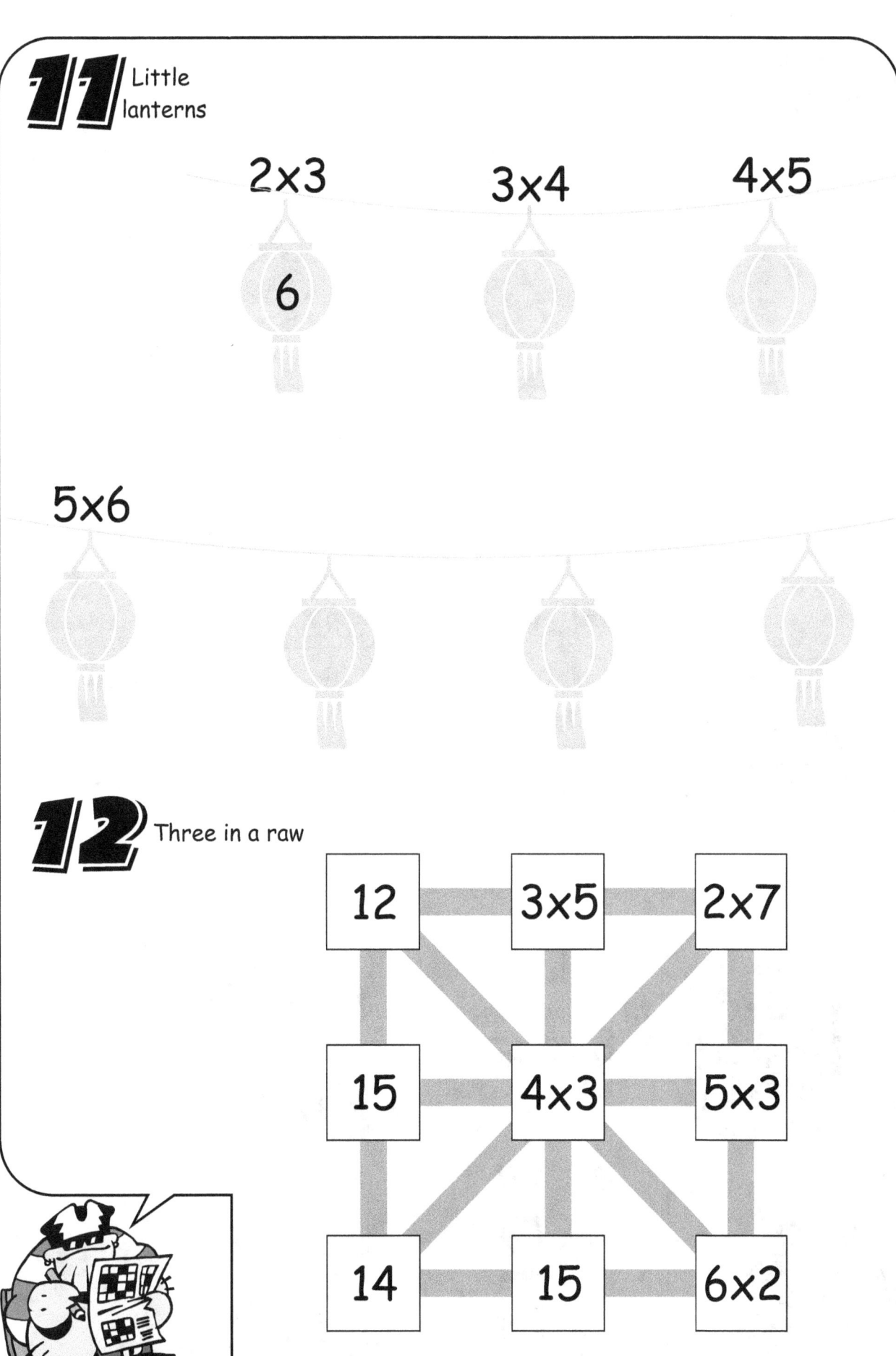

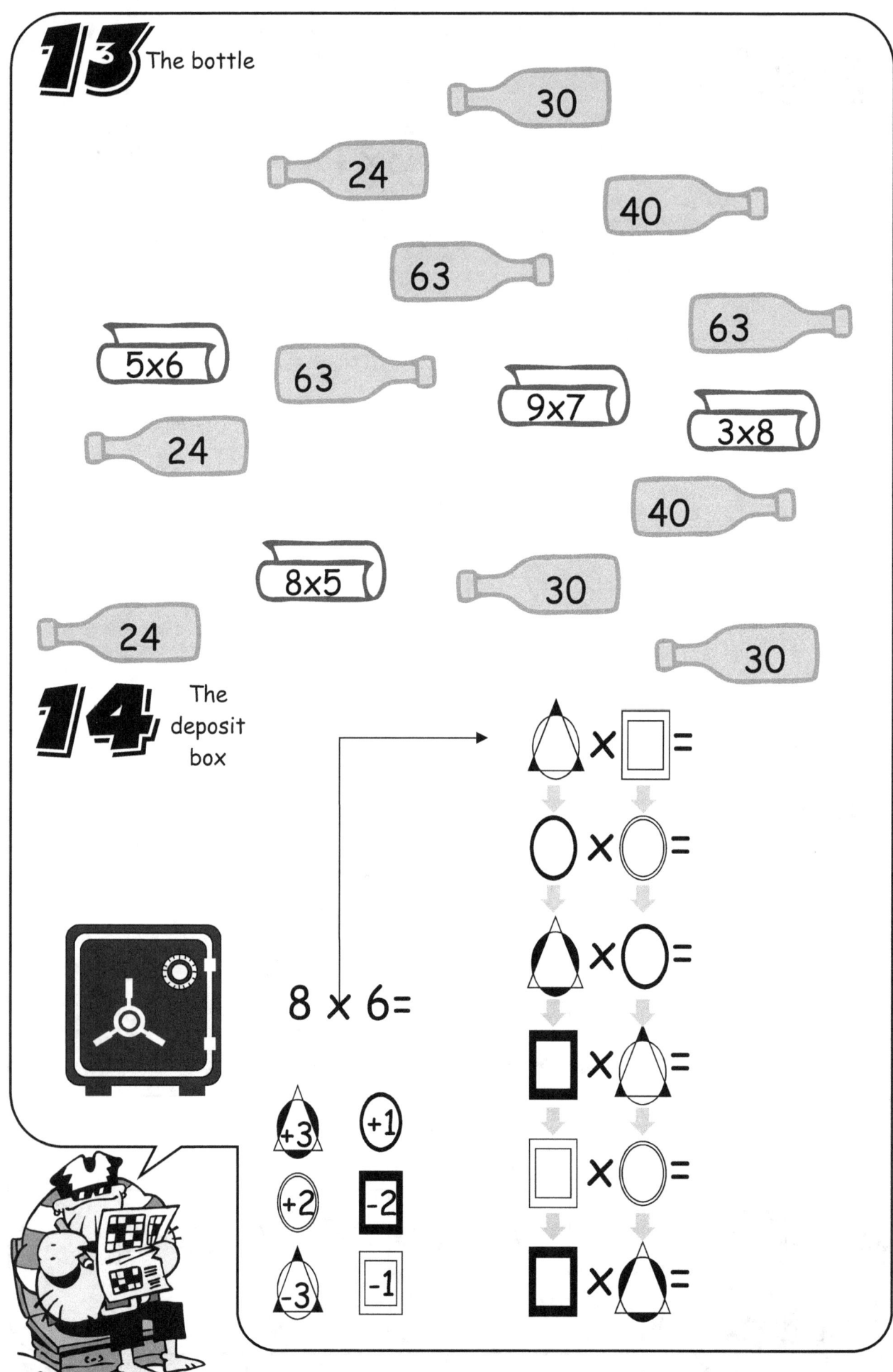

15 The mirror

7 x 7 =
2 x 9 =
3 x 4 =
5 x 6 =
9 x 5 =
8 x 7 =
3 x 9 =
4 x 9 =
9 x 9 =
8 x 9 =
3 x 8 =
5 x 4 =

16 The keys

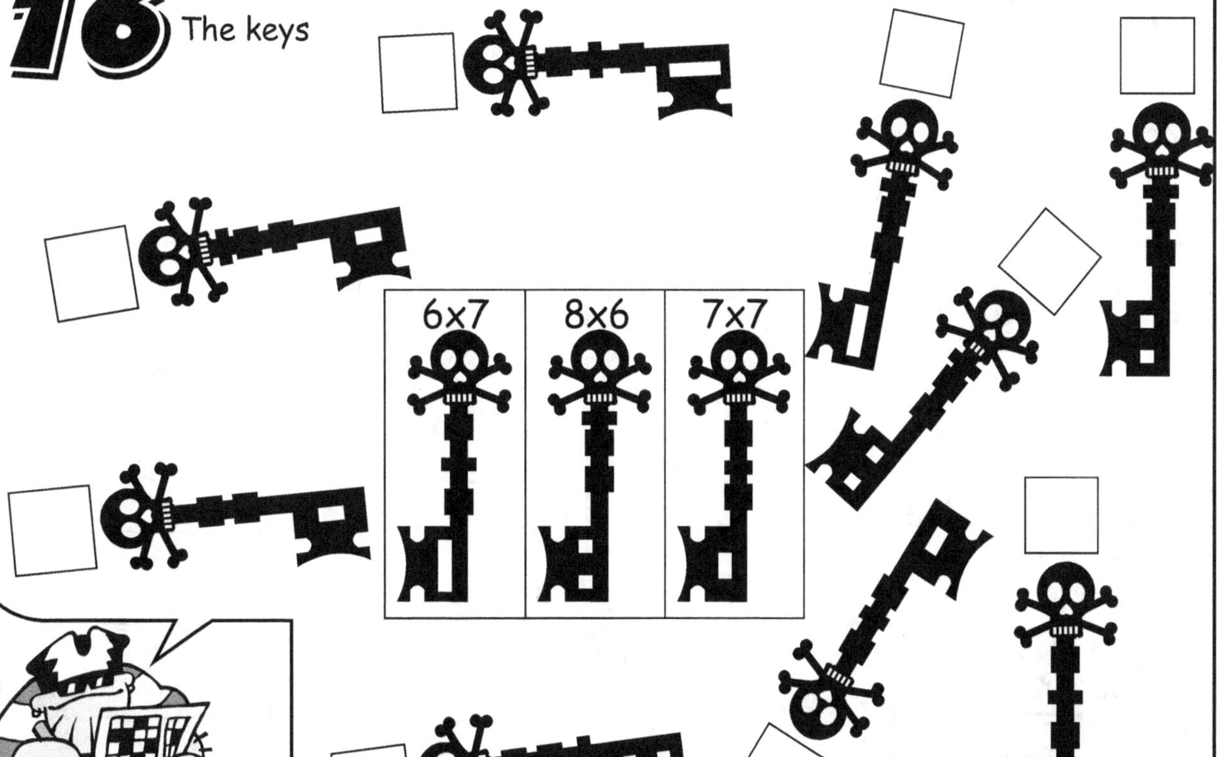

6x7   8x6   7x7

17 The exam

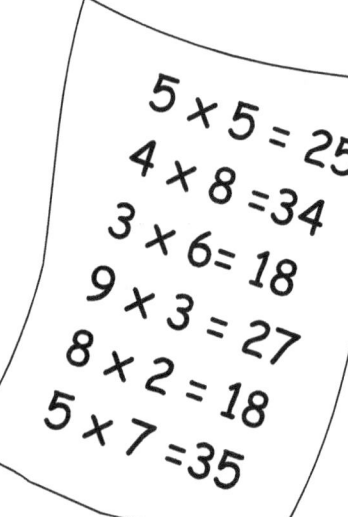

5 × 5 = 25
4 × 8 = 34
3 × 6 = 18
9 × 3 = 27
8 × 2 = 18
5 × 7 = 35

7 × 7 = 48
8 × 7 = 56
6 × 4 = 26
3 × 9 = 25
4 × 9 = 36
9 × 9 = 81

5 × 3 = 15
6 × 2 = 14
7 × 9 = 63
3 × 7 = 22
4 × 4 = 17
8 × 3 = 23

18 Patterms ×

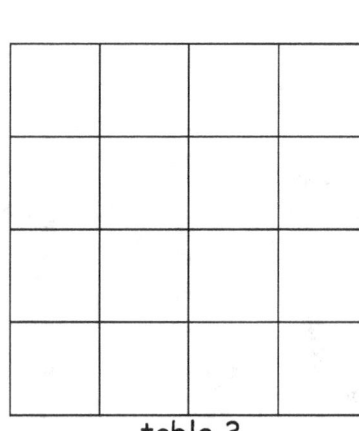

table ¿?=___

table 3

table 4

table 5

19 The paths

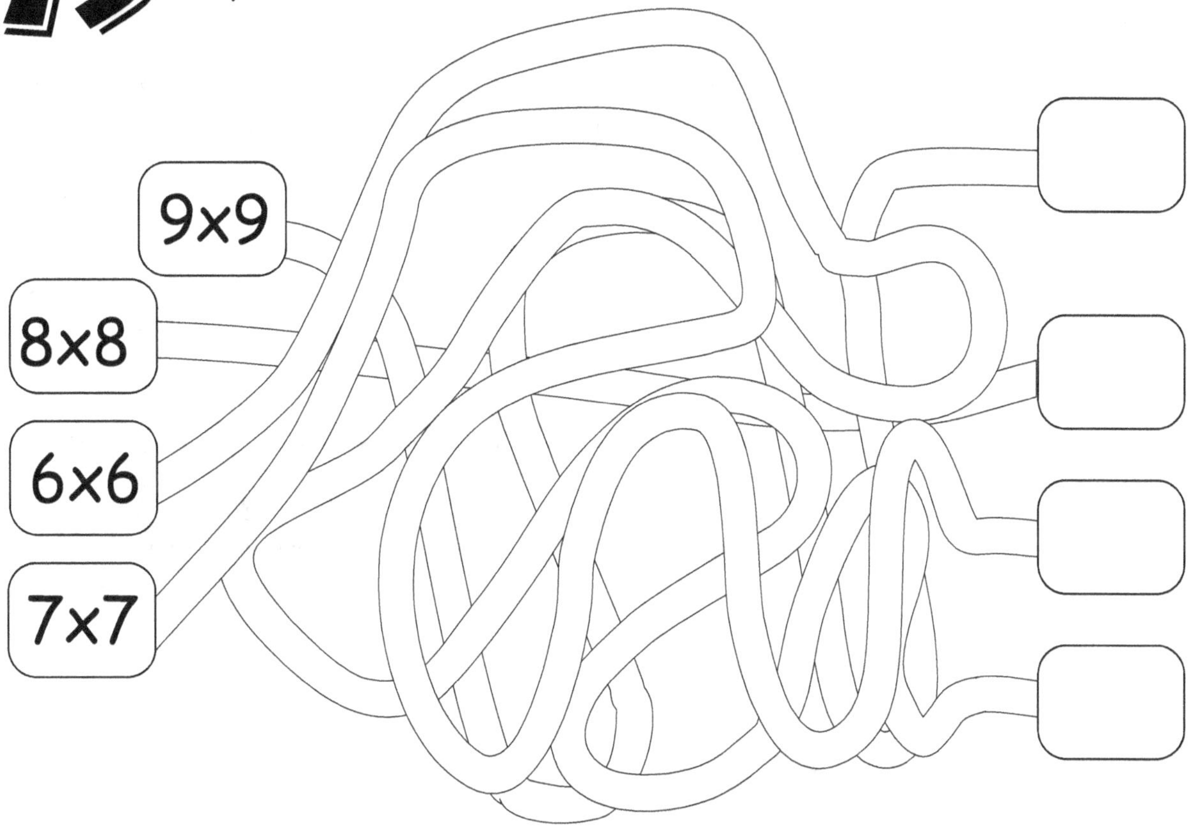

9×9
8×8
6×6
7×7

20 The bomb

21 Domino game

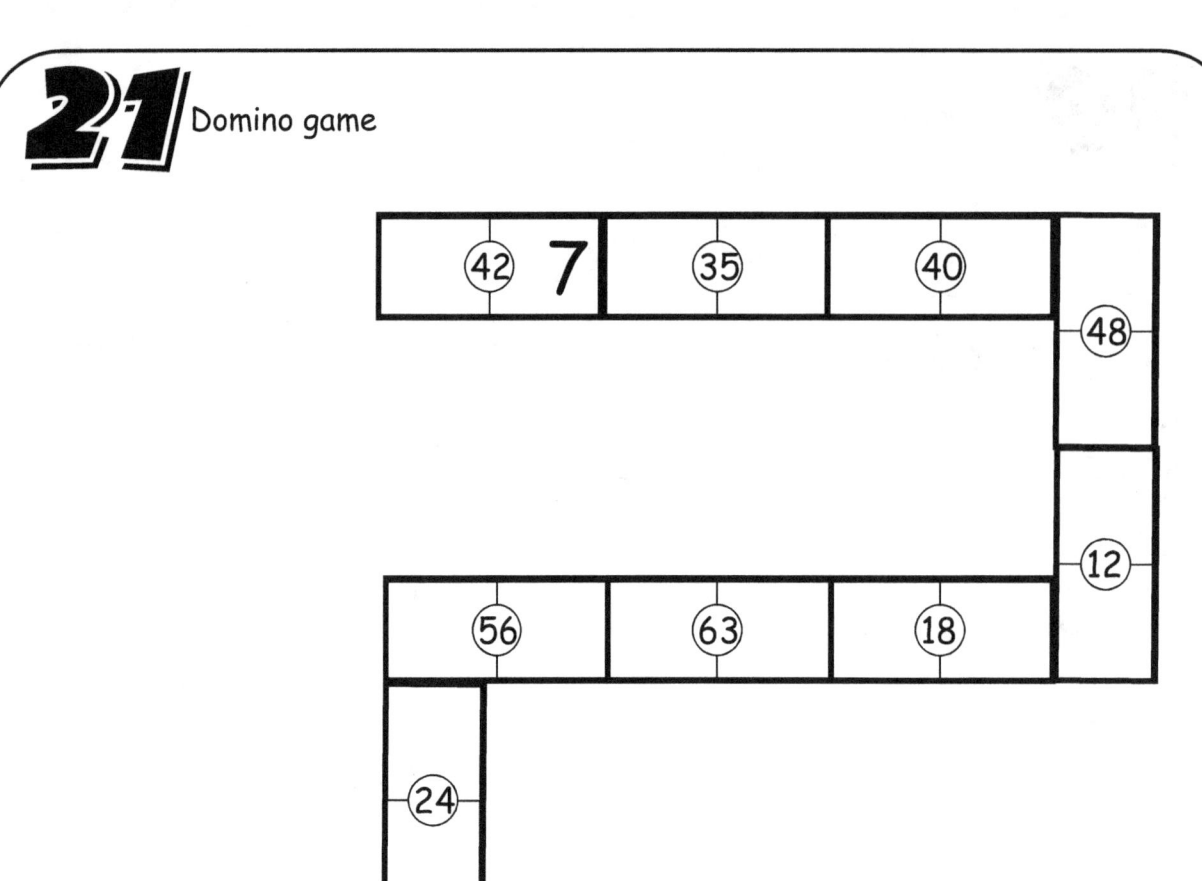

22 The horse

23 Symmetries

6•5	8•4	9•2	4•4	7•9	4•5
4•2	6•6	7•7	8•9	7•4	9•9
6•3	4•9	8•8	5•8	7•2	8•7
6•8	9•4	3•8	4•3	5•4	6•4

24 Investigating

$2 \times \square = 1\square$
$4 \times 6 = \square\square$
$+$
$3\square = 6 \times \square$

$5 \times \square = 1\square$
$3 \times 9 = \square\square$
$+$
$4\square = 6 \times \square$

$4 \times \square = \square 0$
$5 \times \square = 25$
$+$
$4\square = 5 \times \square$

 Diamonds

Guess how many gold coins you will get in the spoils.

	💎	💎	💎	💎
🪙	8	6	5	4
there are				

The value achieved with
Each type of diamond will be:

In this treasure I will be able to get _____ gold coins.

26 Routes

A 8×4
B 4×7
C 7×7
D 8×8
E 9×8

27 Numbers Serarch Puzzle

8	5	45	7	5
4	6	7	42	3
36	4	2	6	14
9	24	3	8	22
4	7	7	48	50

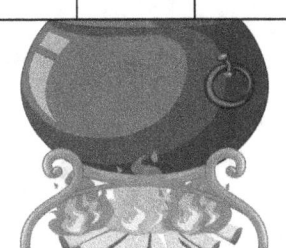

28 the treasure

1.- It is in the 3 and 6 times tables.
2.- It is > than 12.
3.- It is in the 8 times table.

18

24   6   27   12   21

29 The ladder

4x4=

3x3=

4x4=
8x8=
9x8=
3x3=
5x3=
6x7=

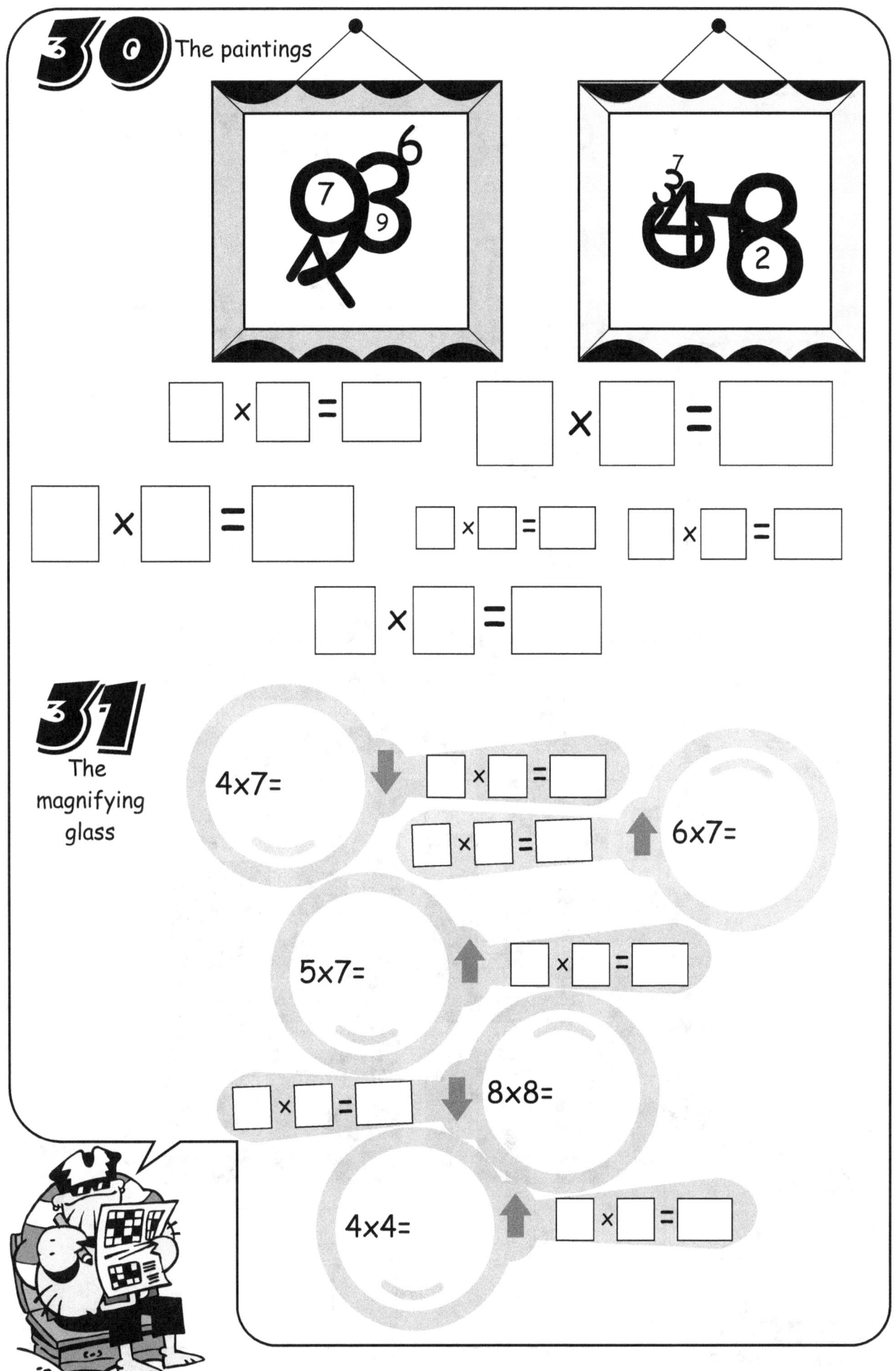

32 The barrels

3x9, 6x7, 2x8,   4x6, 7x8, 9x5

33 The maze

→ Mountain

→ Forest

 "Boxing"

9	6	8
3	4	5
6	8	6

2	5	7
7	7	9
7	8	8

5	4	7
4	7	6
8	9	7

6	8	2
9	3	4
5	7	4

 The swing

1, 2, 3, 10, =

36 Little lanterns

10x1   9x2   8x3

7x4

37 Three in a raw

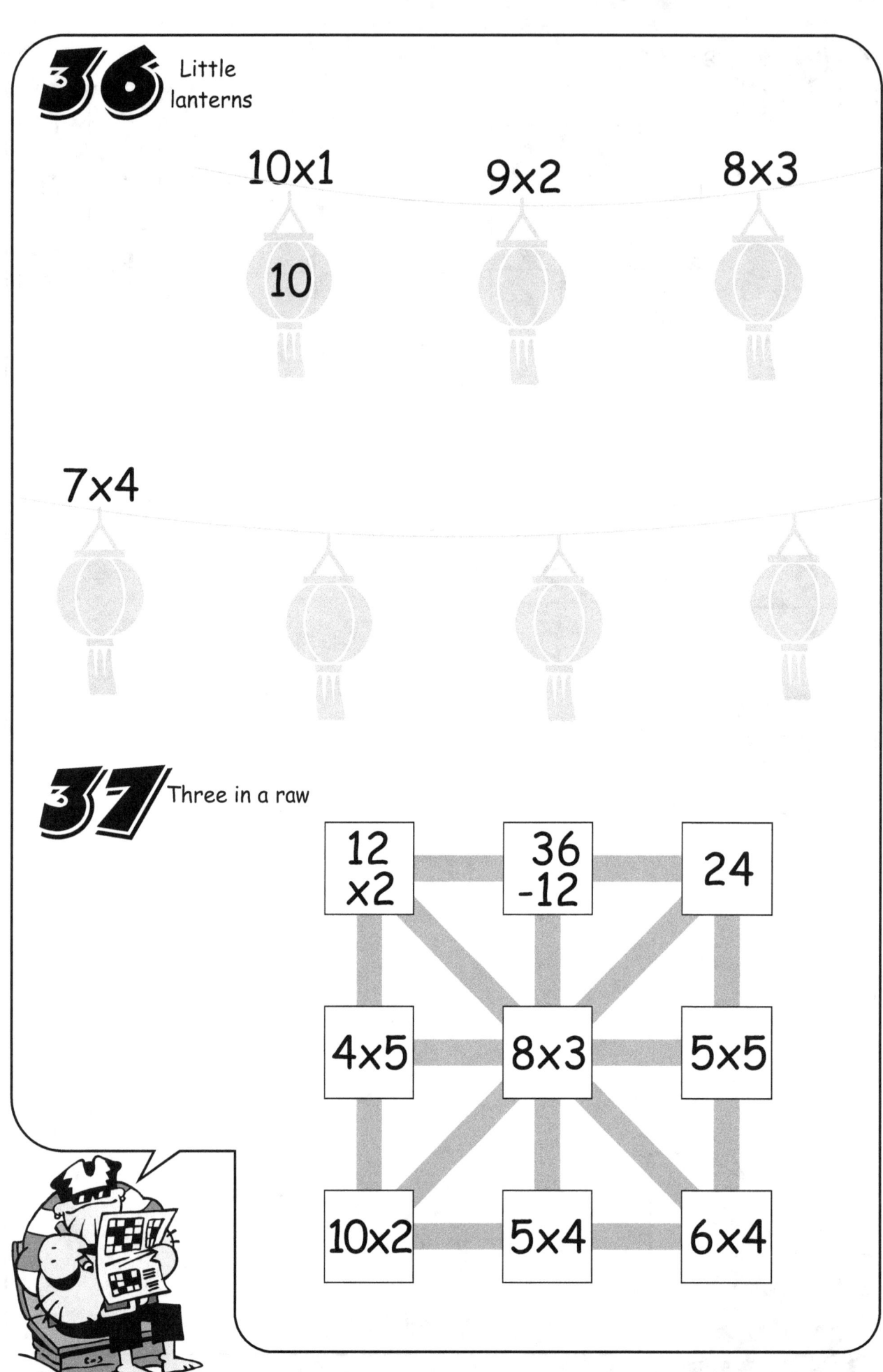

12 x2	36 -12	24
4x5	8x3	5x5
10x2	5x4	6x4

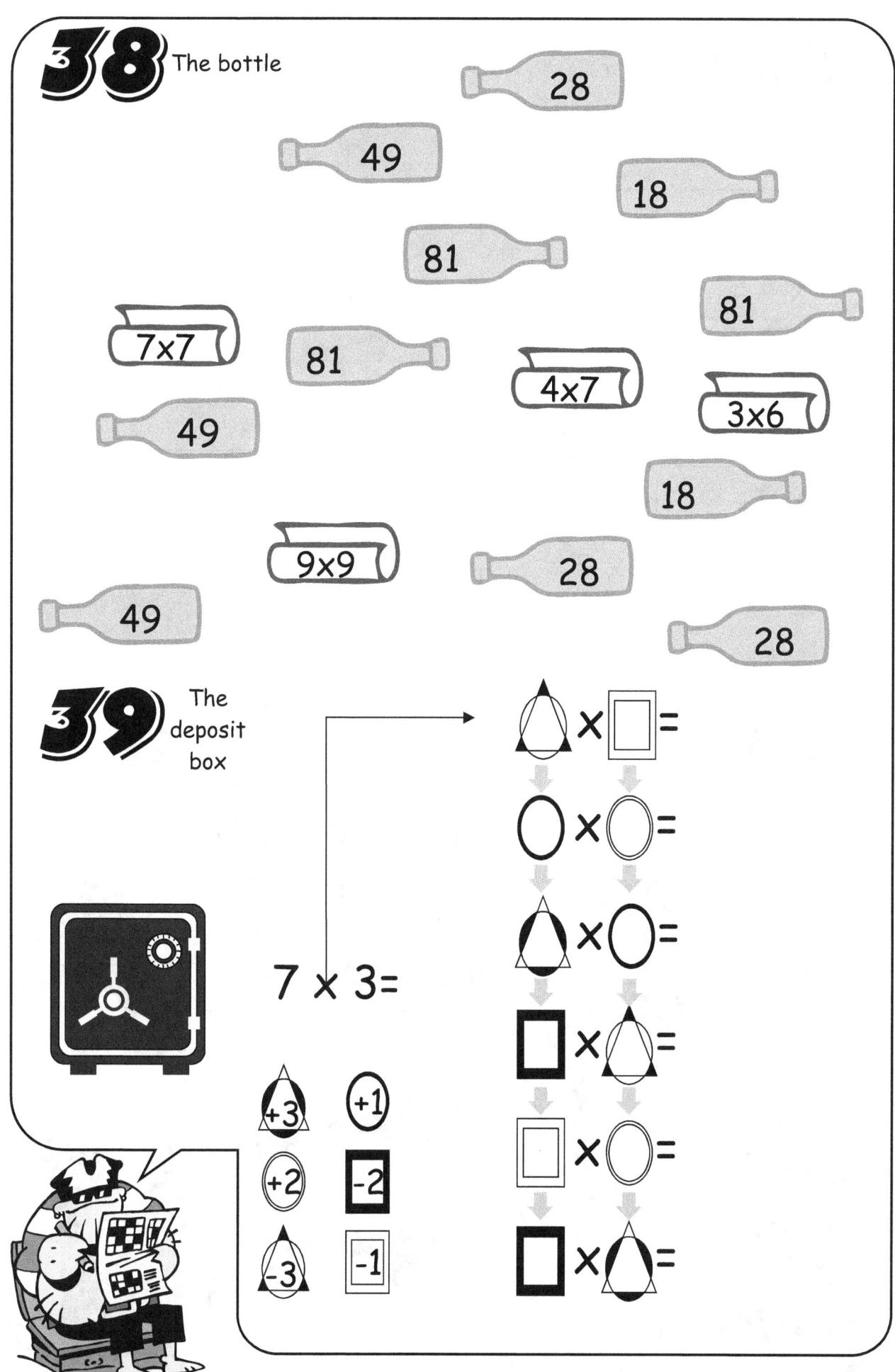

40 The mirror

6 x 6 =   5 x 5 =
4 x 4 =   9 x 9 =
2 x 9 =   7 x 7 =
7 x 2 =   6 x 8 =
6 x 9 =   6 x 4 =
5 x 8 =   5 x 9 =

41 The keys

8x8   9x7   7x4

 The exam

8 × 7 = 56
8 × 8 = 64
7 × 5 = 30
4 × 9 = 34
3 × 3 = 9
2 × 9 = 19

7 × 5 = 35
3 × 8 = 24
5 × 8 = 35
6 × 4 = 24
9 × 3 = 27
2 × 7 = 14

6 × 3 = 16
8 × 2 = 14
9 × 9 = 99
5 × 6 = 40
2 × 4 = 17
4 × 3 = 13

 Patterns ×

18 54
        63
    9
  27

table ¿?=___

table 2

table 7

table 8

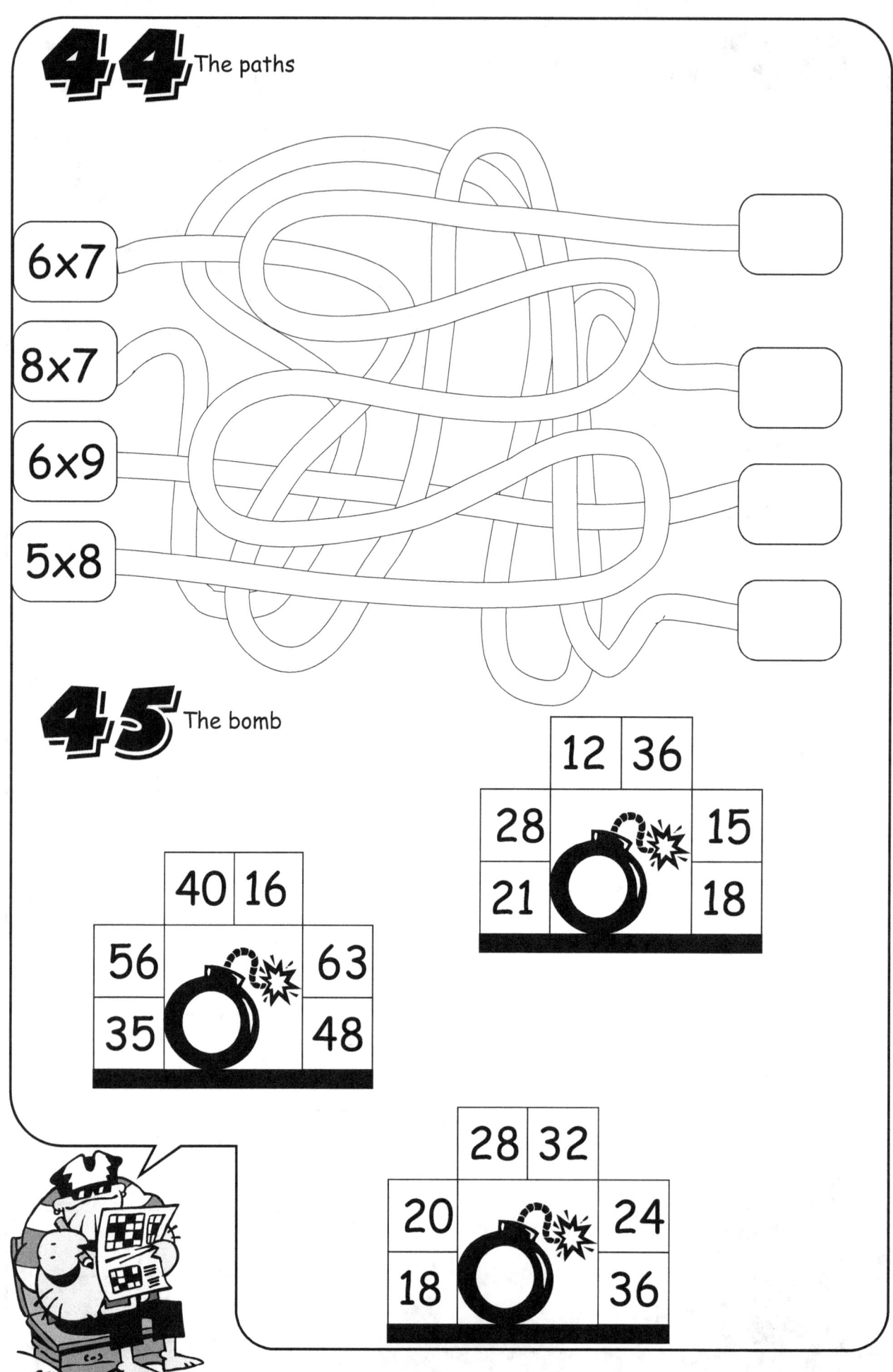

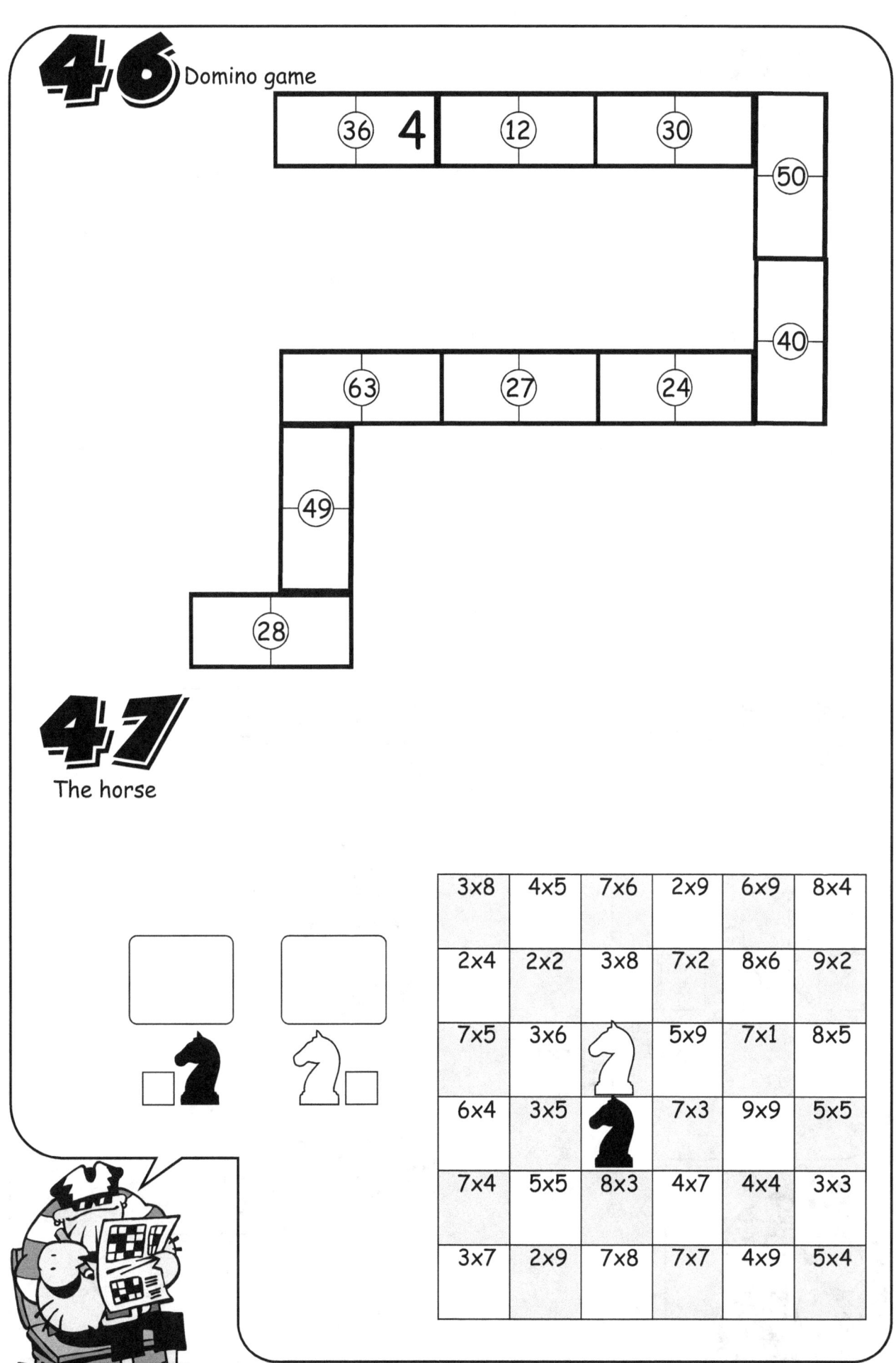

 Symmetries

6×5	8×4	9×6	4×4	7×9	4×5
4×7	6×6	7×7	8×9	7×3	9×9
6×3	4×9	8×8	5×8	7×2	8×7
6×8	9×4	3×8	4×3	5×4	6×4

 Investigating

$5 \times \square = 3\square$
$\square \times 7 = \square 4$
$+$
$\overline{4\square} = 7 \times \square$

$4 \times \square = 3\square$
$6 \times \square = 2\square$
$+$
$\overline{5\square} = \square \times 8$

$6 \times \square = \square 2$
$7 \times \square = 2\square$
$+$
$\overline{6\square} = \square \square$

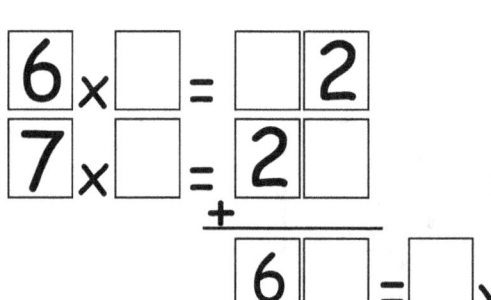

50 Diamonds

Guess how many gold coins you will get in the spoils.

	💎	💎	💎	💎
🪙	9	7	3	6
there are				

The value achieved with Each type of diamond will be:

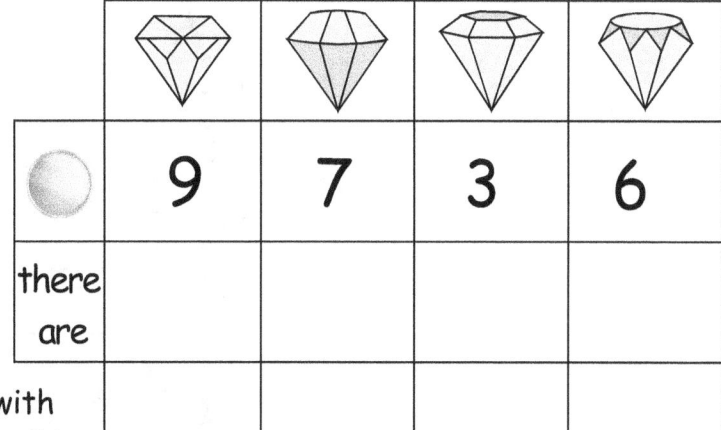

In this treasure I will be able to get _____ gold coins.

-43-

51 Routes

A 48:6
B 36:9
C 63:9
D 24:8
E 27:3

52 Numbers Serarch Puzzle

9	4	27	3	36
6	8	56	4	6
49	7	7	21	6
24	6	8	48	1
17	42	7	5	1

 the treasure

1.- Its quotient is number 9.
2.- Its Dividend is an odd number.
3.- Double its dividend contains a 1 a 2 and a 6 number.
4.- Double its dividend is >130.

72:8

15:5   81:9   63:7   27:3   40:5

54 The ladder

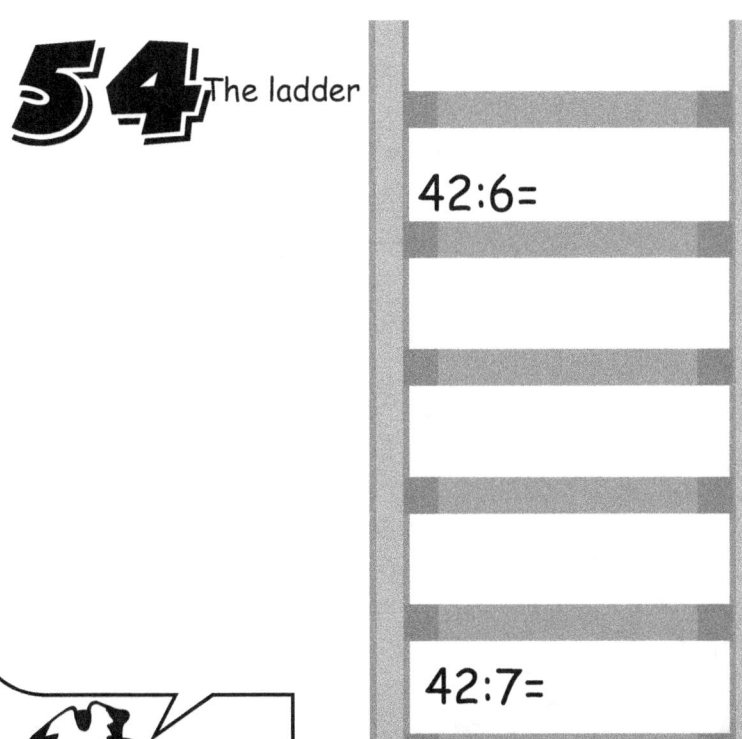

42:6=

42:7=

42:6
42:7
81:9
80:8
35:7
15:3

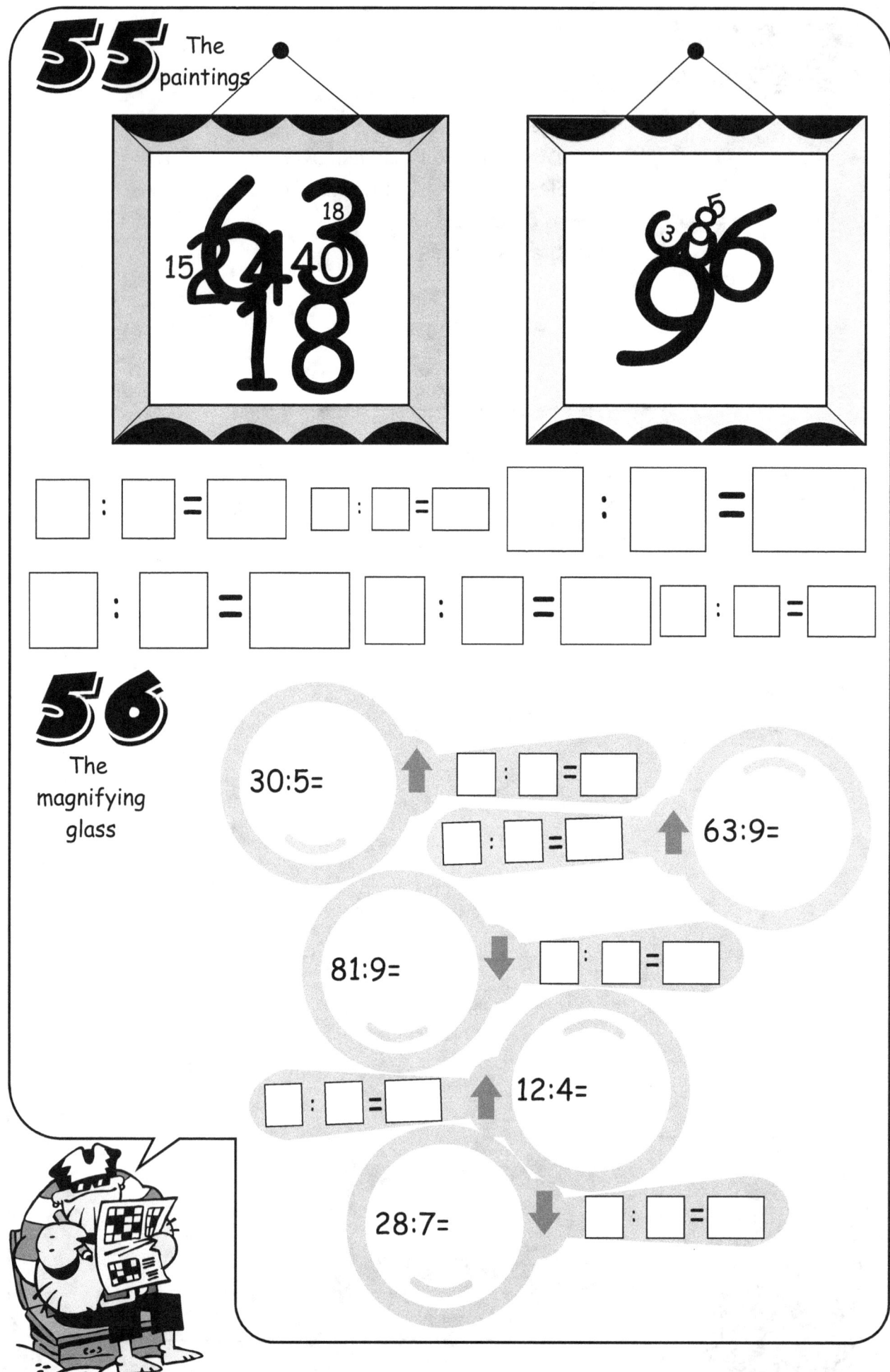

57
The barrels

21:7, 32:4, 45:9, 28:7, 63:9, 18:2

58
The maze

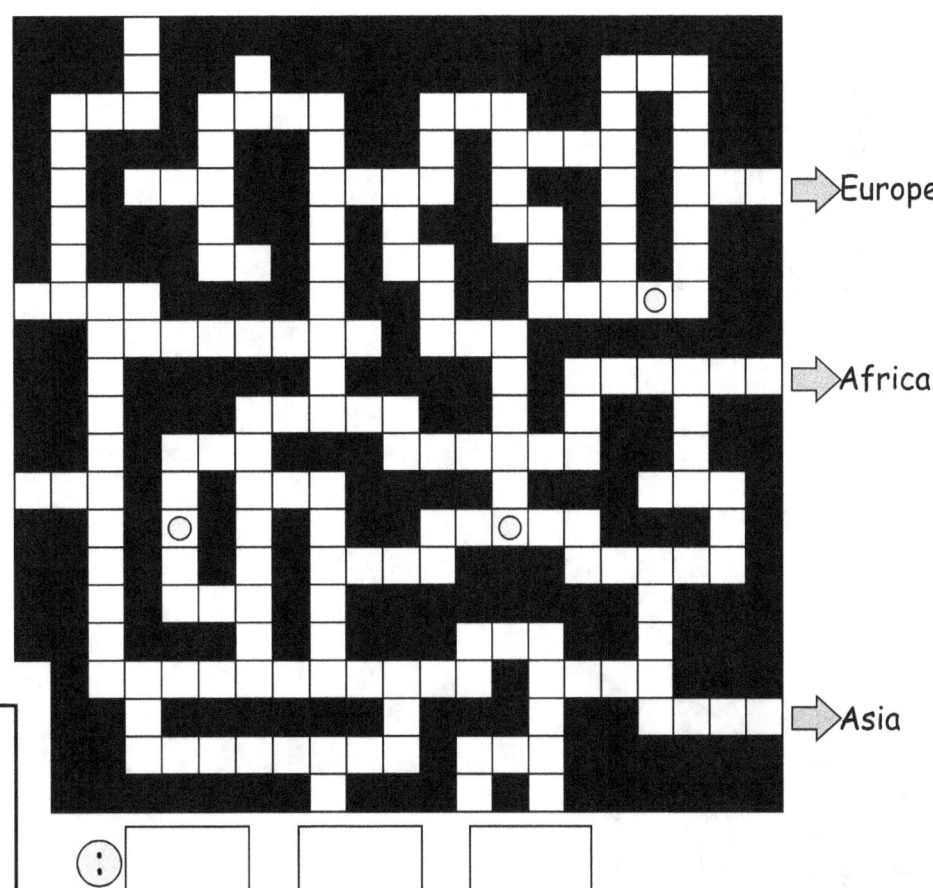

→ Europe

→ Africa

→ Asia

"Boxing"

64	20	27
12	15	42
45	28	30

8	5	9
6	3	6
5	4	5

4	9	3
7	2	8
4	6	9

16	18	21
42	8	40
36	18	54

The swing

1, 2, 3, 10, =

 Little lanterns

18:2   27:3   36:4

9

45:5

62 Three in a raw

49:7	42:6	63:9
21:3	18:3	24:4
56:8	48:8	36:6

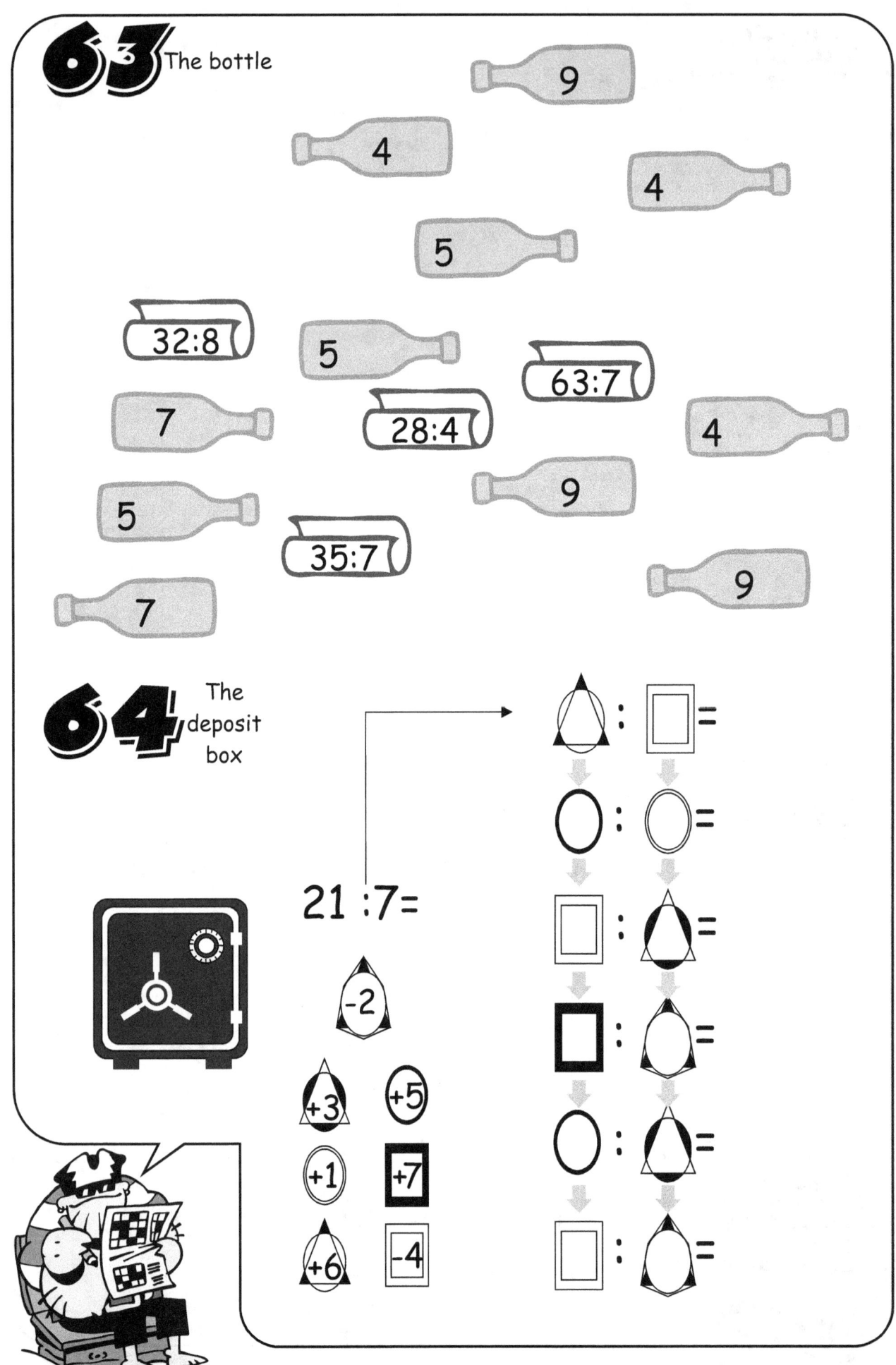

65 The mirror

8 × 8 =
9 × 9 =
6 × 6 =
2 × 6 =
8 × 9 =
6 × 5 =
3 × 3 =
7 × 7 =
7 × 3 =
5 × 3 =
4 × 7 =
9 × 7 =

66 The keys

8×3 9×4 6×9

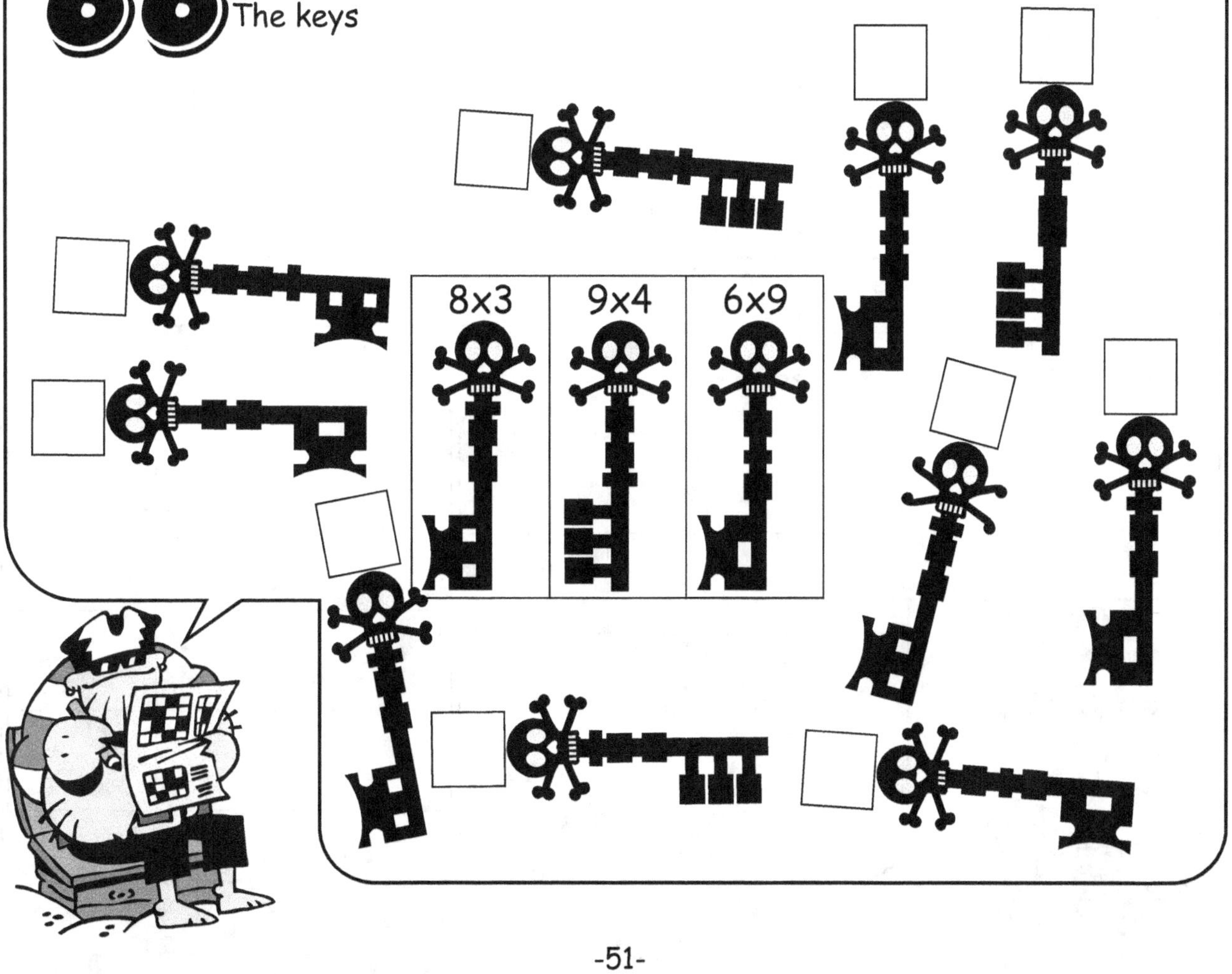

 The exam

81 : 9 = 8
48 : 6 = 8
24 : 8 = 4
36 : 4 = 8
63 : 7 = 9
18 : 6 = 2

36 : 6 = 5
45 : 5 = 8
72 : 9 = 9
56 : 8 = 8
42 : 6 = 6
64 : 8 = 8

35 : 7 = 5
25 : 5 = 4
27 : 3 = 8
16 : 4 = 3
54 : 9 = 6
81 : 9 = 9

 Patterms

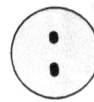

Result 8

Result 6

Result 7

Result 4

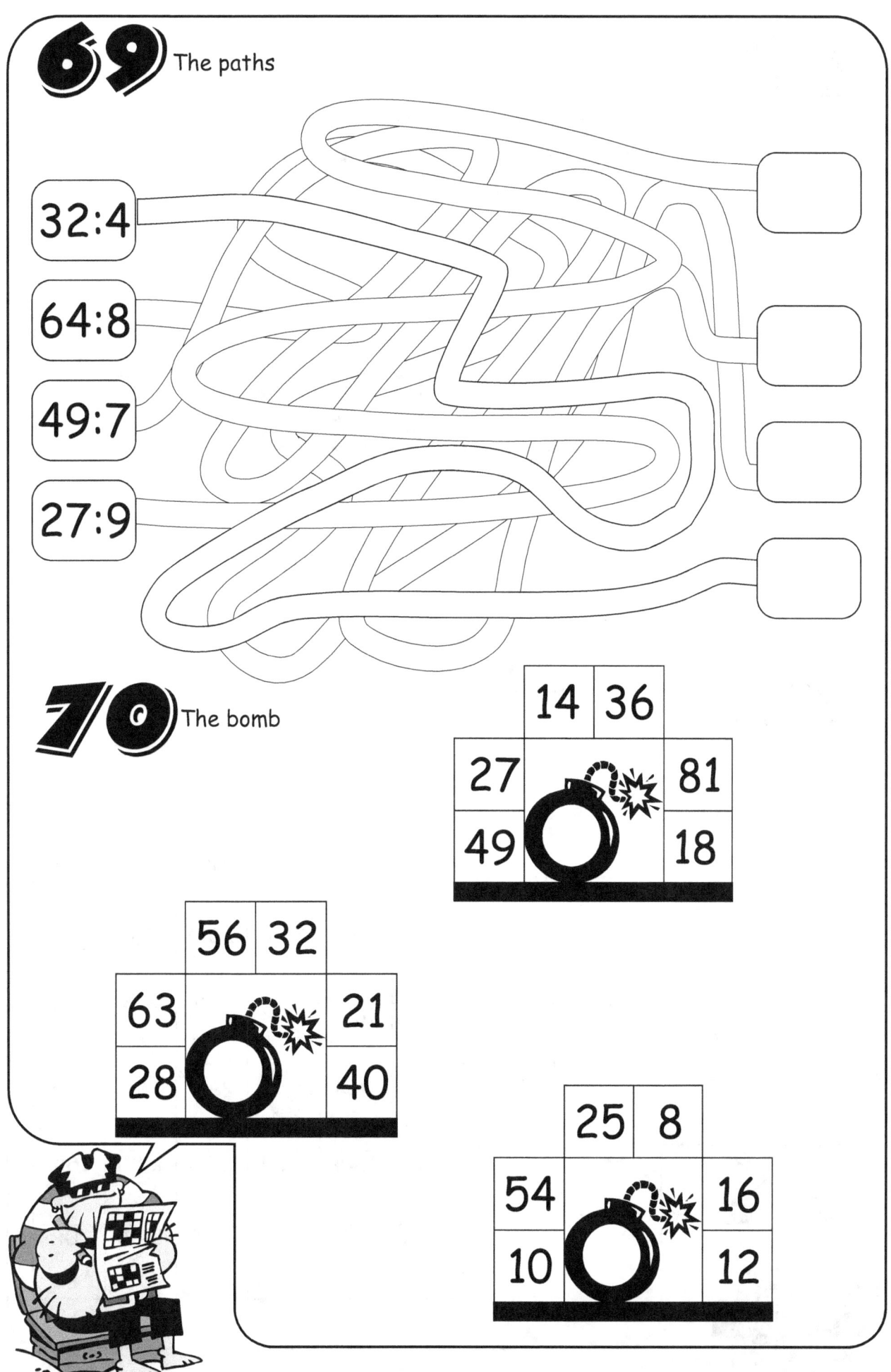

71 Domino game

72 The horse

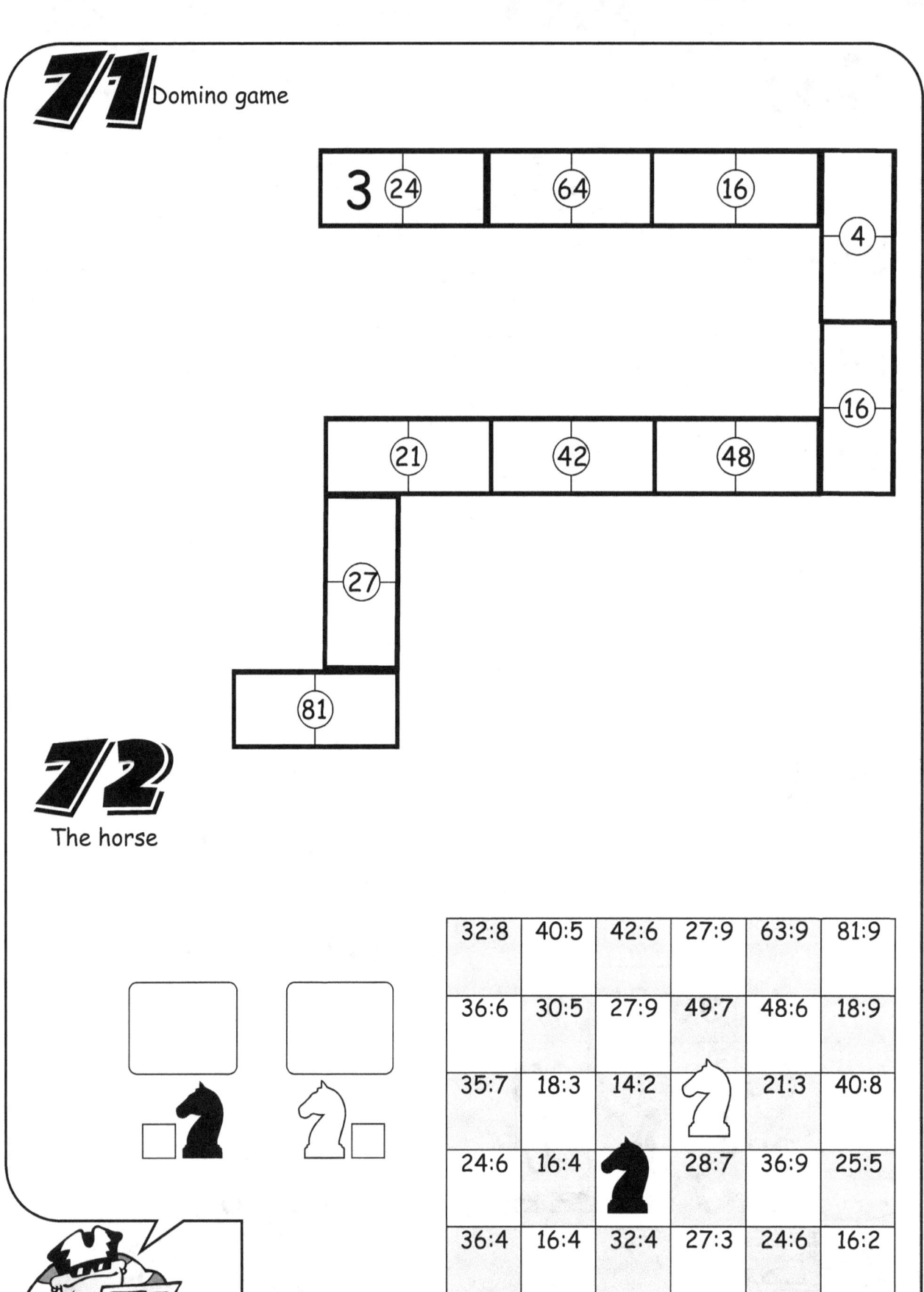

 Symmetries

30●5	81●9	36●6	35●7	72●8	30●5
32●4	24●6	20●2	16●8	70●7	81●9
64●8	18●9	54●6	42●6	15●3	25●5
21●7	28●7	36●9	27●3	40●5	56●8

 Investigating

$21 : \square = 3$
$\square : 4 = \square$
$+$
$45 : \square = \square$
$+\_\_\_$
16

$35 : \square = 5$
$\square : 3 = 5$
$+$
$50 : \square = \square$
$+\_\_\_$
15

$48 : \square = 8$
$\square : \square = \square$
$+$
$63 : \square = 9$
$+\_\_\_$
16

-55-

75 Diamonds

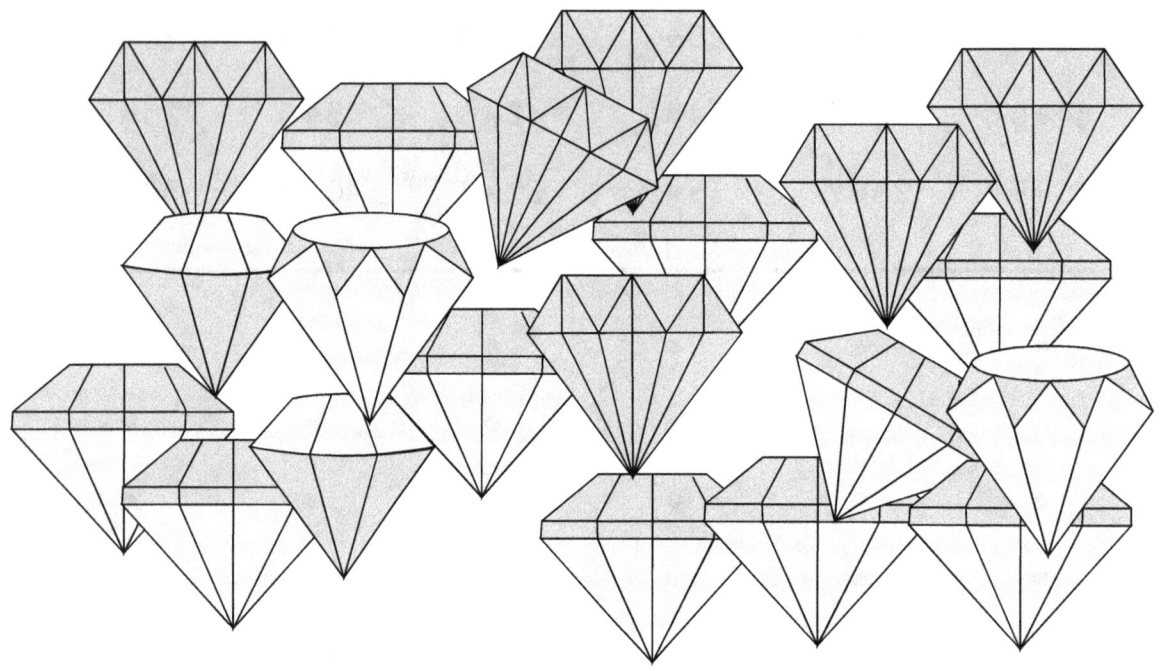

Guess how many gold coins you will get in the spoils.

	💎	💎	💎	💎
🪙	8	6	2	9
there are				
If we are to distribute...	4	9	4	6
Each of us will touch ...				

The value achieved with each type of diamond will be:

76 Routes

A 21:7
B 25:5
C 18:2
D 32:4
E 49:7

77 Numbers Serarch Puzzle

72	45	28	4	7
9	3	18	2	8
8	2	36	2	4
8	9	6	54	3
64	9	6	3	8

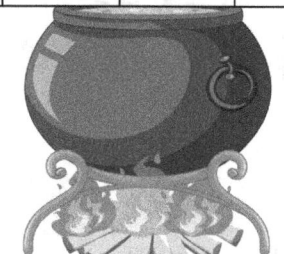

78 the treasure

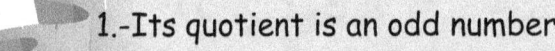

1.- Its quotient is an odd number.
2.- Its dividend is not in the 10 times table.
3.- Its quotient is three times the rest.

24:8

28:4   48:8   56:7   40:8   36:4

79 The ladder

56:8=

63:9=

56:8
63:9
49:7
25:5
40:8
20:2

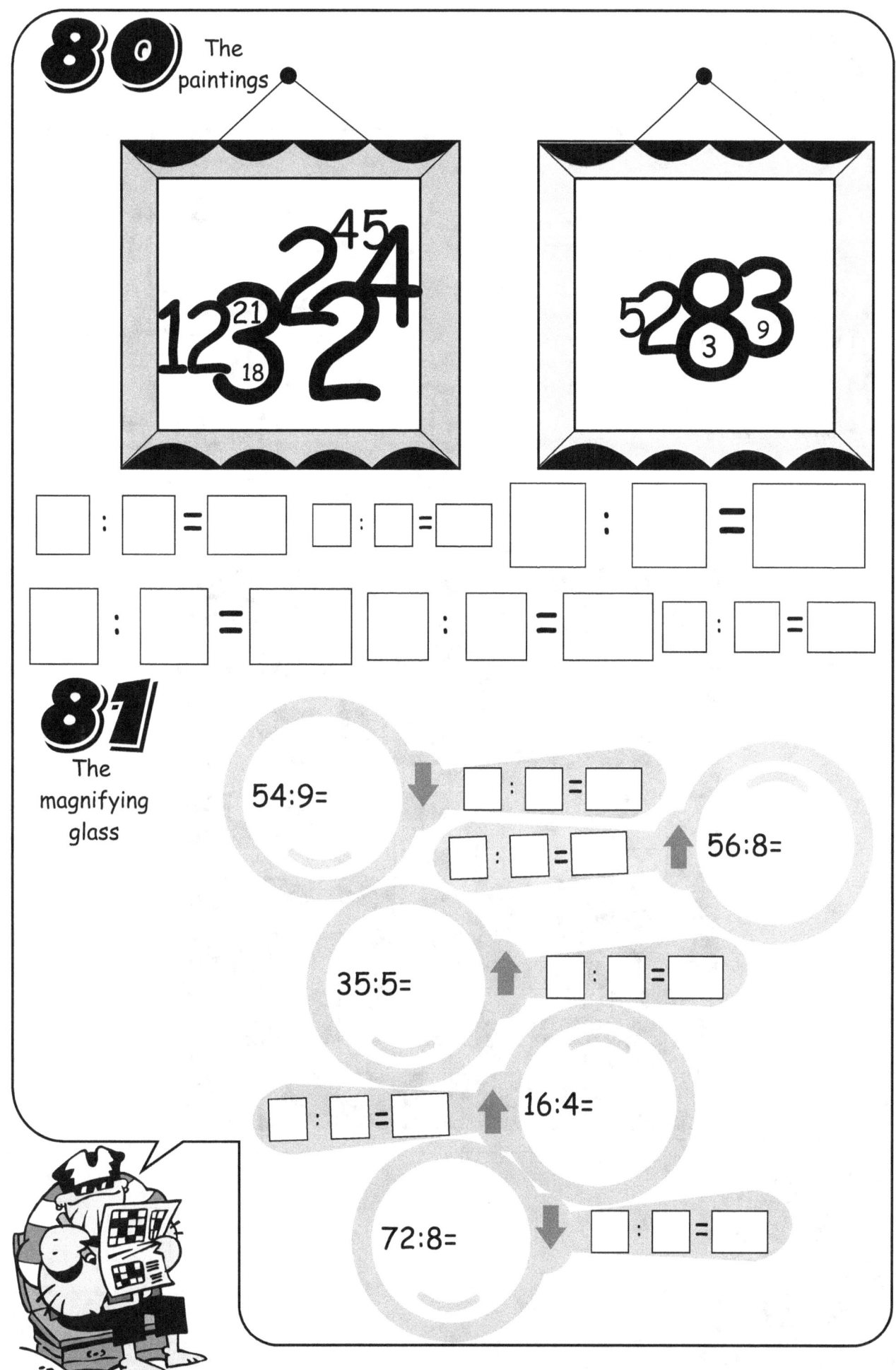

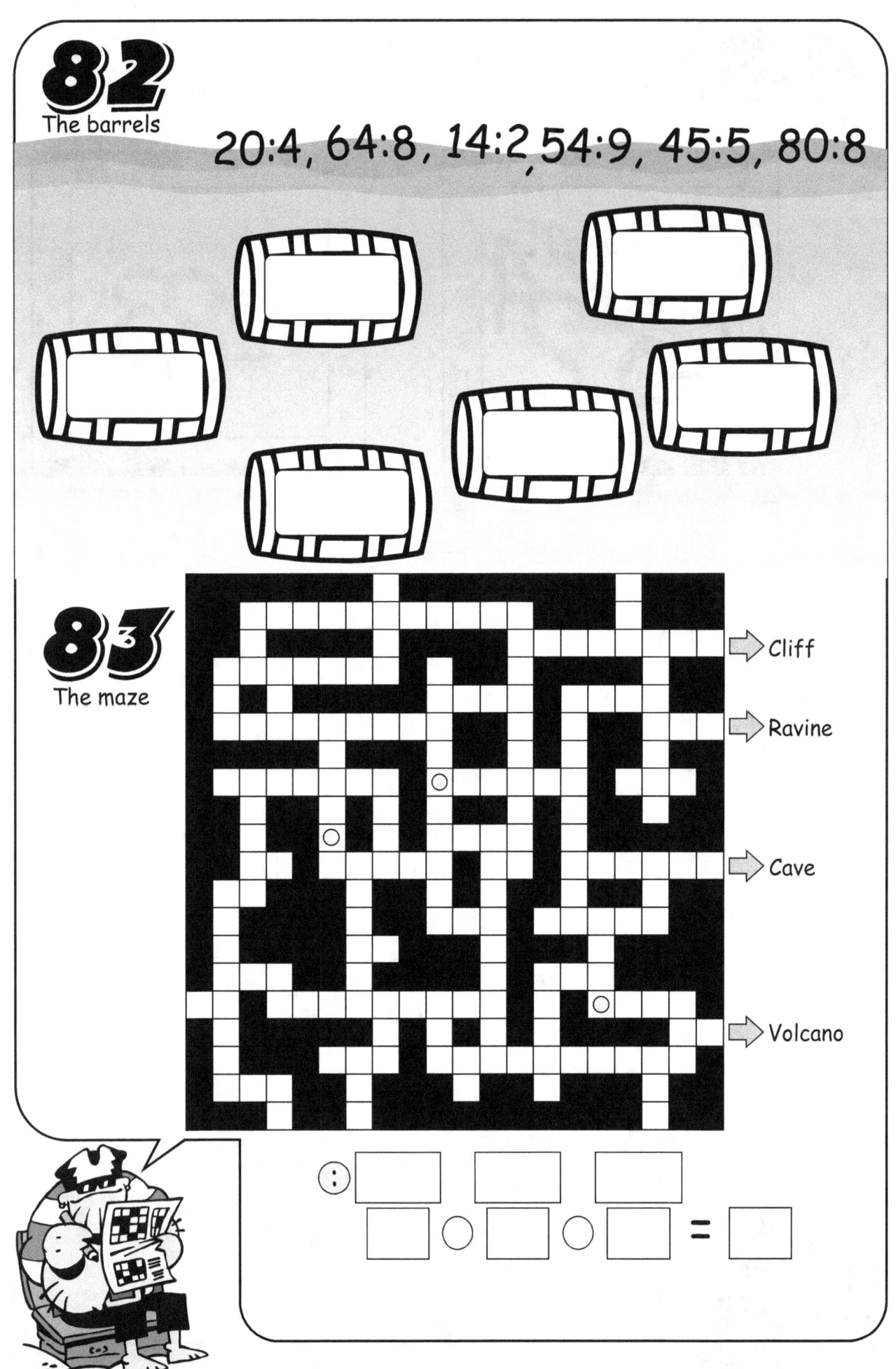

84 "Boxing"

14	81	36
36	24	49
72	16	7

3	7	7
8	8	8
4	8	6

15	28	63
32	56	64
12	16	60

7	9	6
9	8	7
9	8	7

85 The swing

1, 2, 3, 10, =

86 Little lanterns

60:10   54:9   48:8

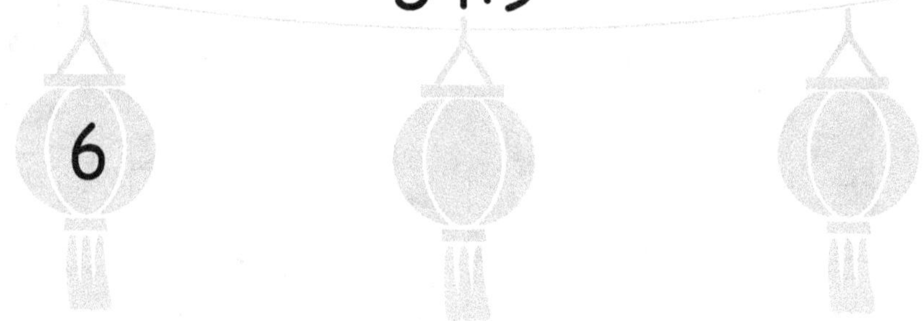

6

42:7

87 Three in a raw

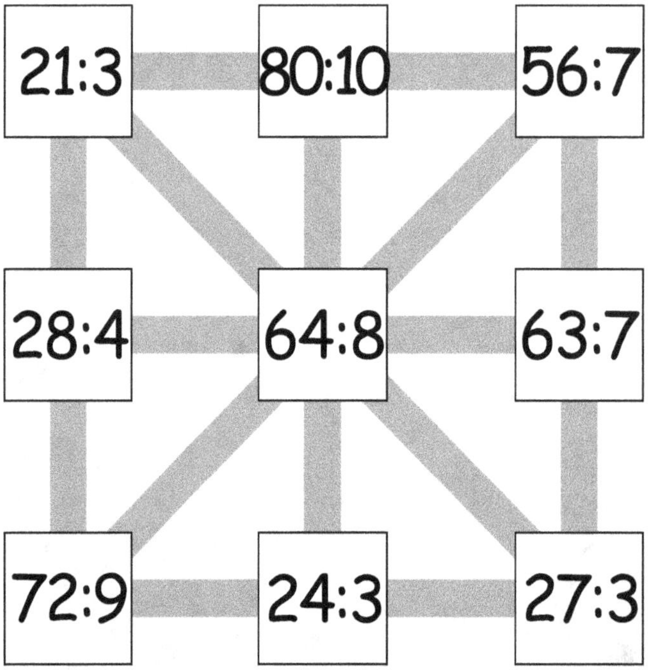

21:3	80:10	56:7
28:4	64:8	63:7
72:9	24:3	27:3

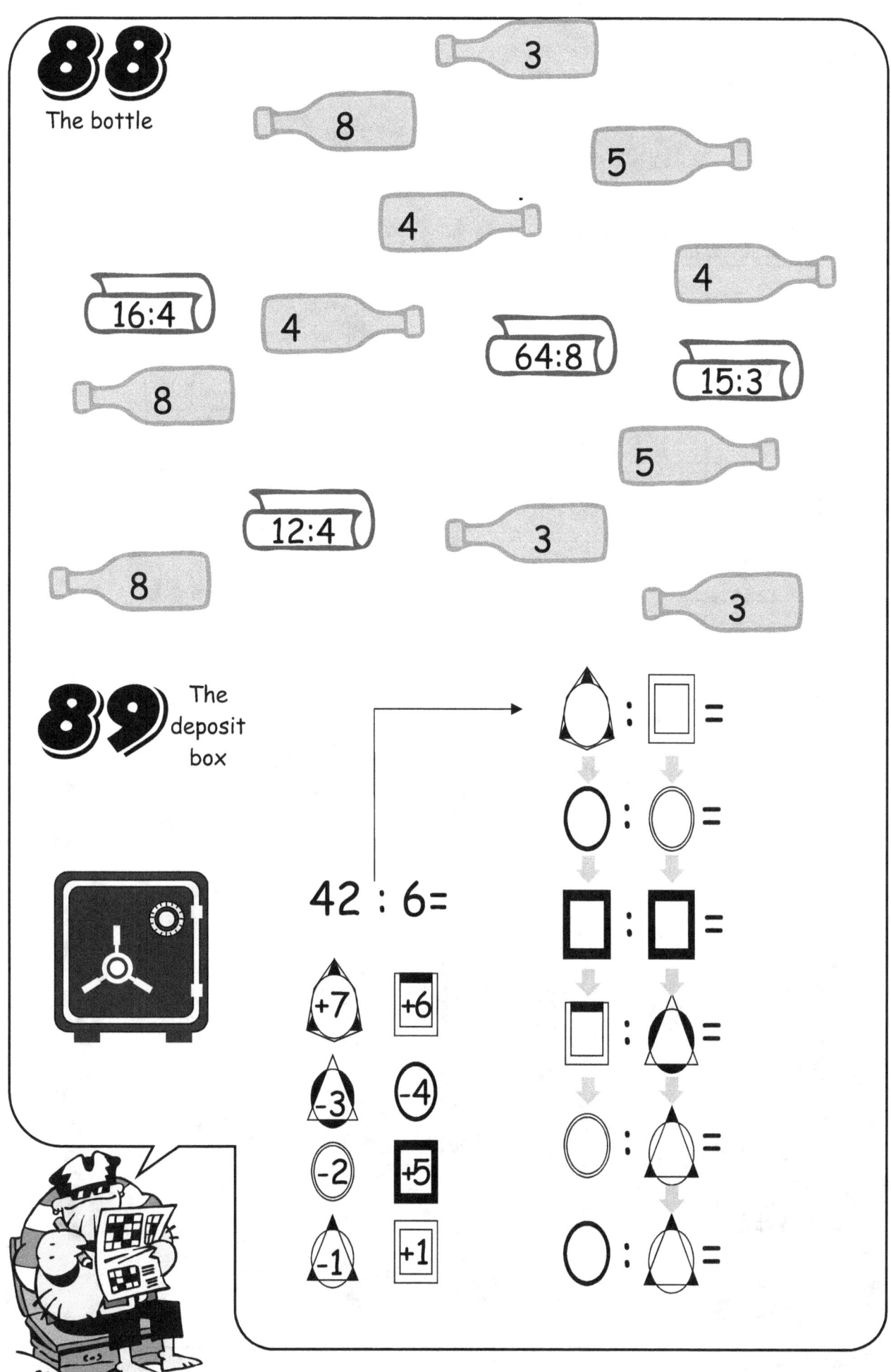

90 The mirror

8 × 9 =
2 × 9 =
4 × 2 =
5 × 6 =
3 × 6 =
6 × 7 =
5 × 9 =
4 × 3 =
8 × 3 =
7 × 5 =
10 × 4 =
3 × 9 =

91 The keys

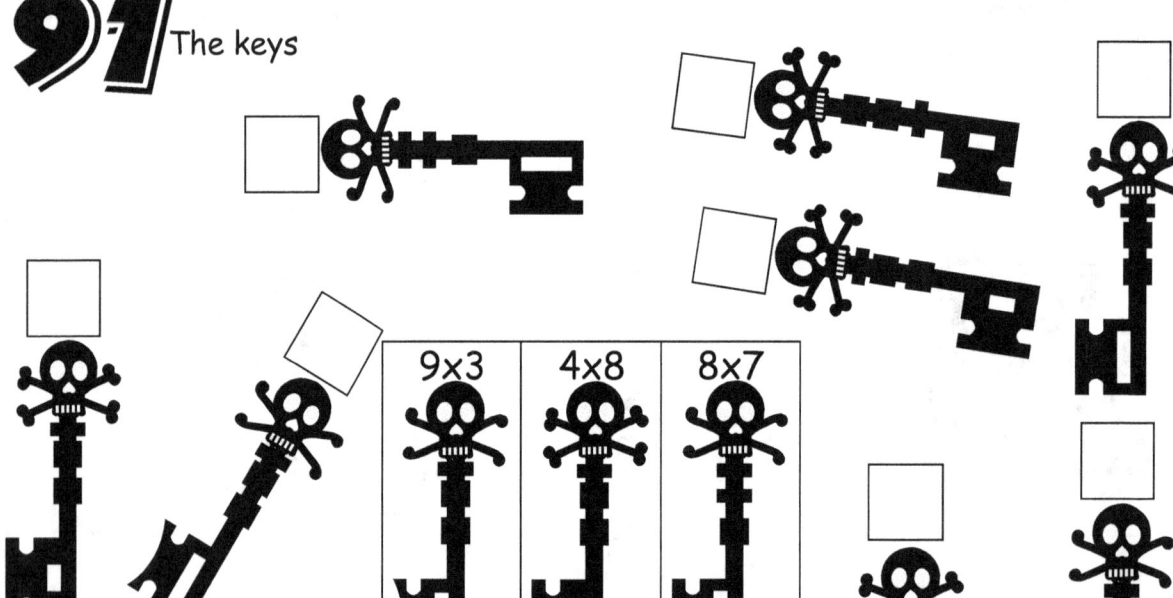

 The exam

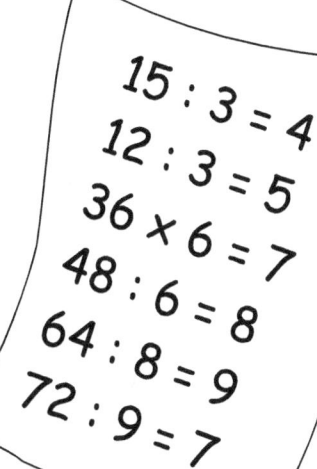

15 : 3 = 4
12 : 3 = 5
36 x 6 = 7
48 : 6 = 8
64 : 8 = 9
72 : 9 = 7

14 : 7 = 3
50 : 5 = 10
18 : 2 = 9
32 : 8 = 4
36 : 6 = 7
28 : 4 = 6

21 : 3 = 6
14 : 7 = 3
63 : 9 = 7
21 : 7 = 3
16 : 4 = 4
24 : 8 = 3

 Patterms

Result 9

Result 5

Result 3

Result 7

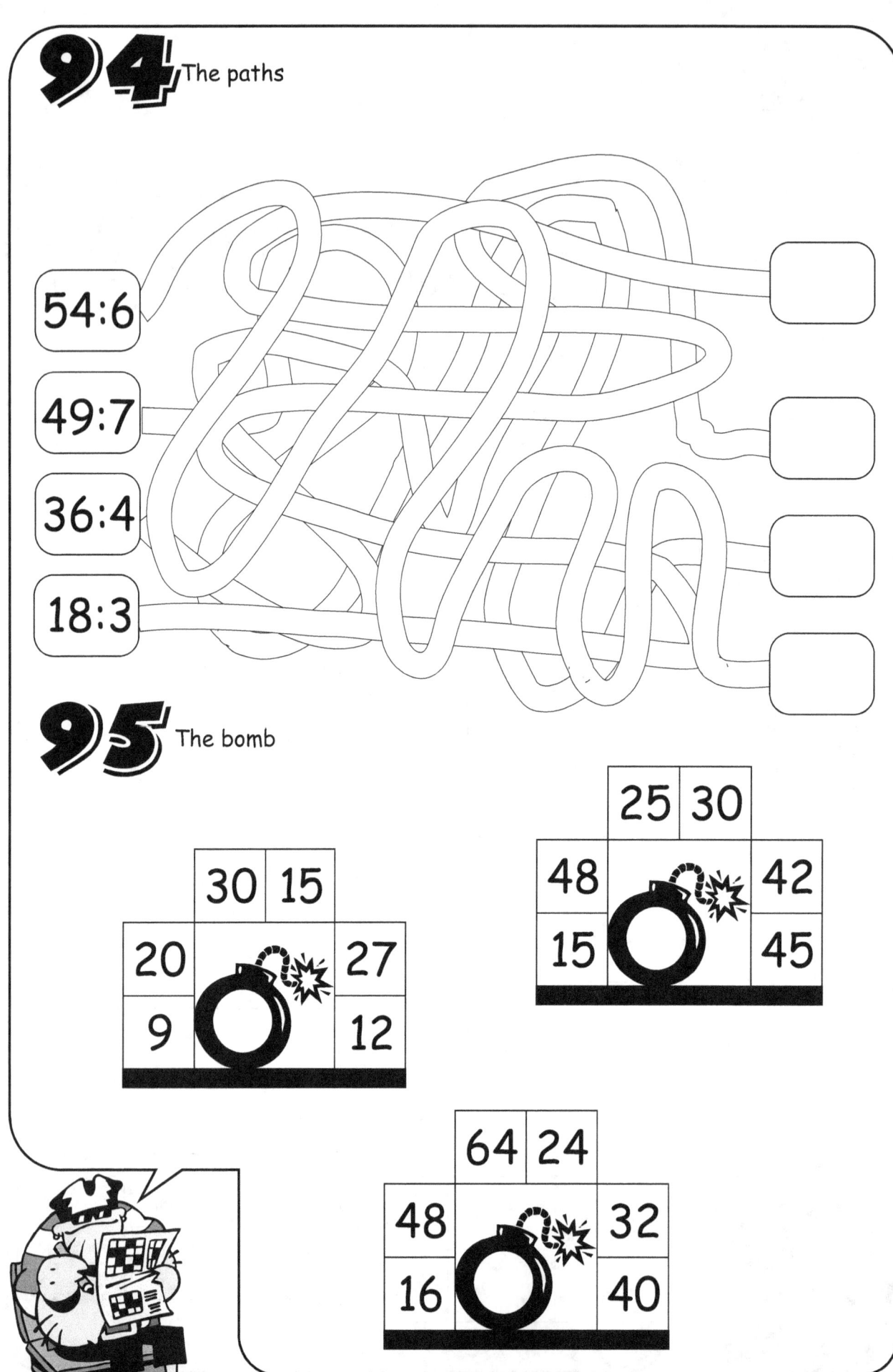

96 Domino game

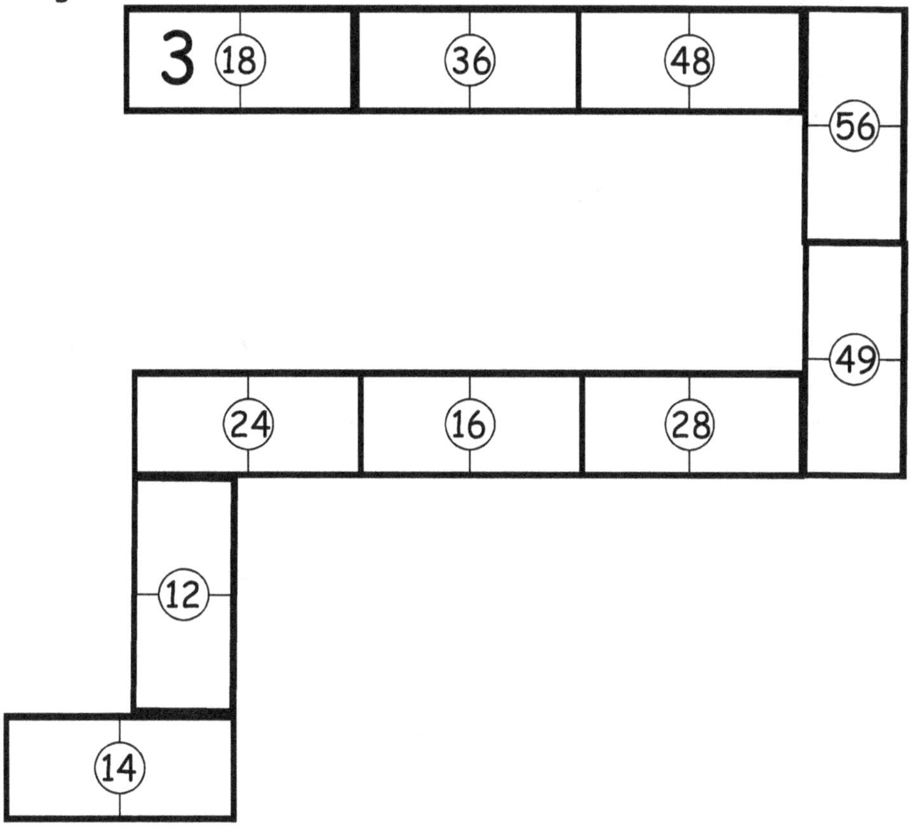

3 ⑱	㊱	㊽	
			㊺
			㊾
㉔	⑯	㉘	
⑫			
⑭			

97 The horse

32:8	40:5	42:6	27:9	63:9	81:9
36:6	45:5	27:9	49:7	48:6	18:9
35:7		14:2	27:3	21:3	40:8
24:6	15:5	18:3	28:7	36:9	25:5
36:4	16:4	32:4		24:6	16:2
24:3	6:2	12:4	45:5	54:6	20:5

 Symmetries

30●5	81●9	48●8	35●7	72●8	30●5
64●8	24●6	20●2	16●8	63●7	81●9
64●8	18●9	54●6	42●6	15●3	40●8
21●7	28●7	36●9	56●8	40●5	56●8

 Investigating

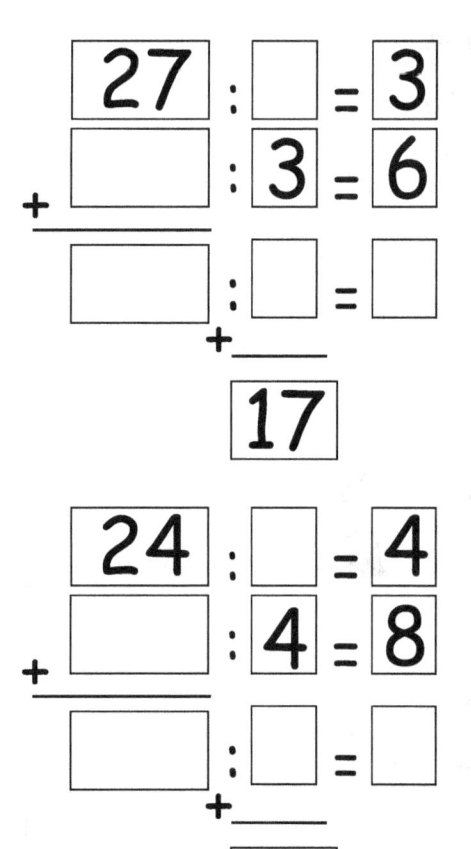

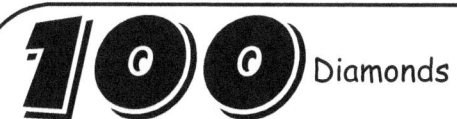 Diamonds

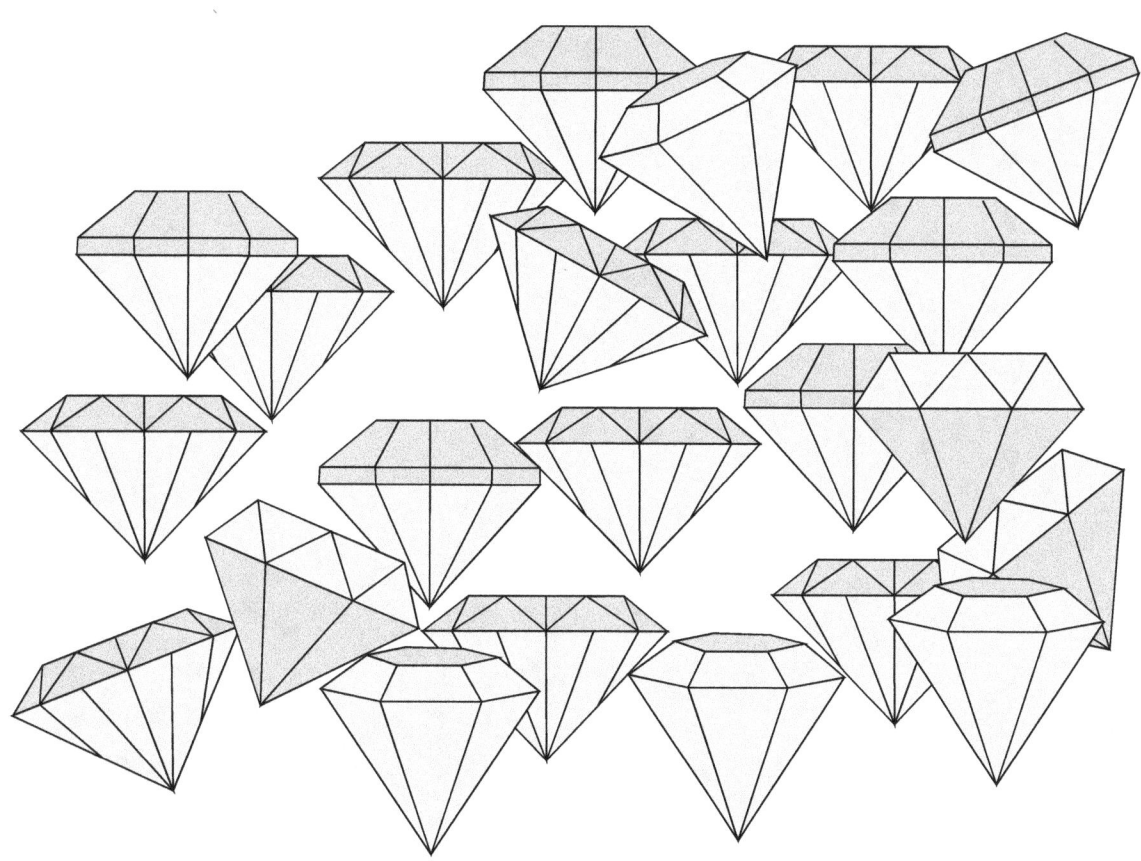

Guess how many gold coins you will get in the spoils.

The value achieved with each type of diamond will be:

	💎	💎	💎	💎
⚪	8	3	4	6
there are				
If we are to distribute...	4	6	8	4
Each of us will touch ...				

-69-

Here at the back you will have a treasure that will help you

ANSWER KEY

The video game download code is redraven

1
36B
35D
21C
24A

2
12	2	6	5	32
2	3	8	3	24
5	3	4	12	5
3	10	34	9	4
15	7	2	5	10

3
56

4
several solutions
2x4=8
3x5=15
8x6=48
9x3=27
5x8=40
3x7=21

5
9x3=27
7x2=14
8x3=24
5x4=20
6x2=12
4x7=28

6
2x5=10 - 4x3=12
3x5=15 - 4x4=16
4x9=36 - 5x7=35
6x4=24 - 5x5=25
7x3=21 - 4x5=20

7
3x7=21
6x4=24
9x3=27
8x4=32
5x7=35
4x9=36

8
Island A
Island B

2x10 | 3x7 | 4x2
20 ⊕ 21 ⊕ 8 = 49

9
16	18	21
42	8	40
36	18	42

56	8	27
12	15	42
45	28	30

11
2x3=6
3x4=12
4x5=20
5x6=30
6x7=42
7x8=56
8x9=72

12
12 - 15 - 14
15 - 12 - 15
14 - 15 - 12

13
30, 24, 40, 63, 63, 5x6, 63, 9x7, 3x8, 24, 40, 8x5, 30, 24, 30

14
8x6=48
5x5=25
6x7=42
9x8=72
7x5=35
6x7=42
4x10=40

15
7x7=49 | 2x9=18
3x4=12 | 5x6=30
9x5=45 | 8x7=56
3x9=27 | 4x9=36
9x9=81 | 8x9=72
3x8=24 | 5x4=20

-72-

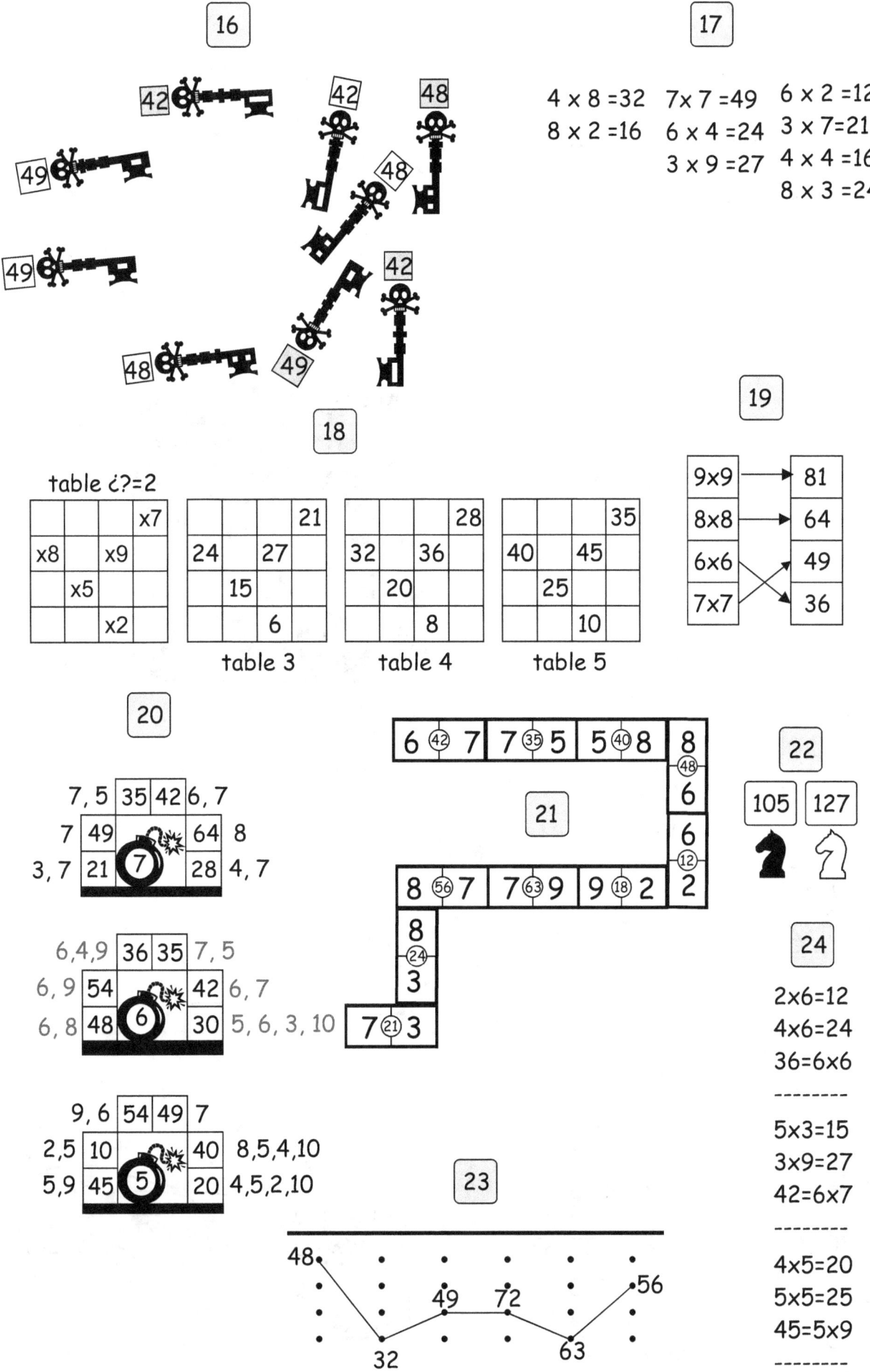

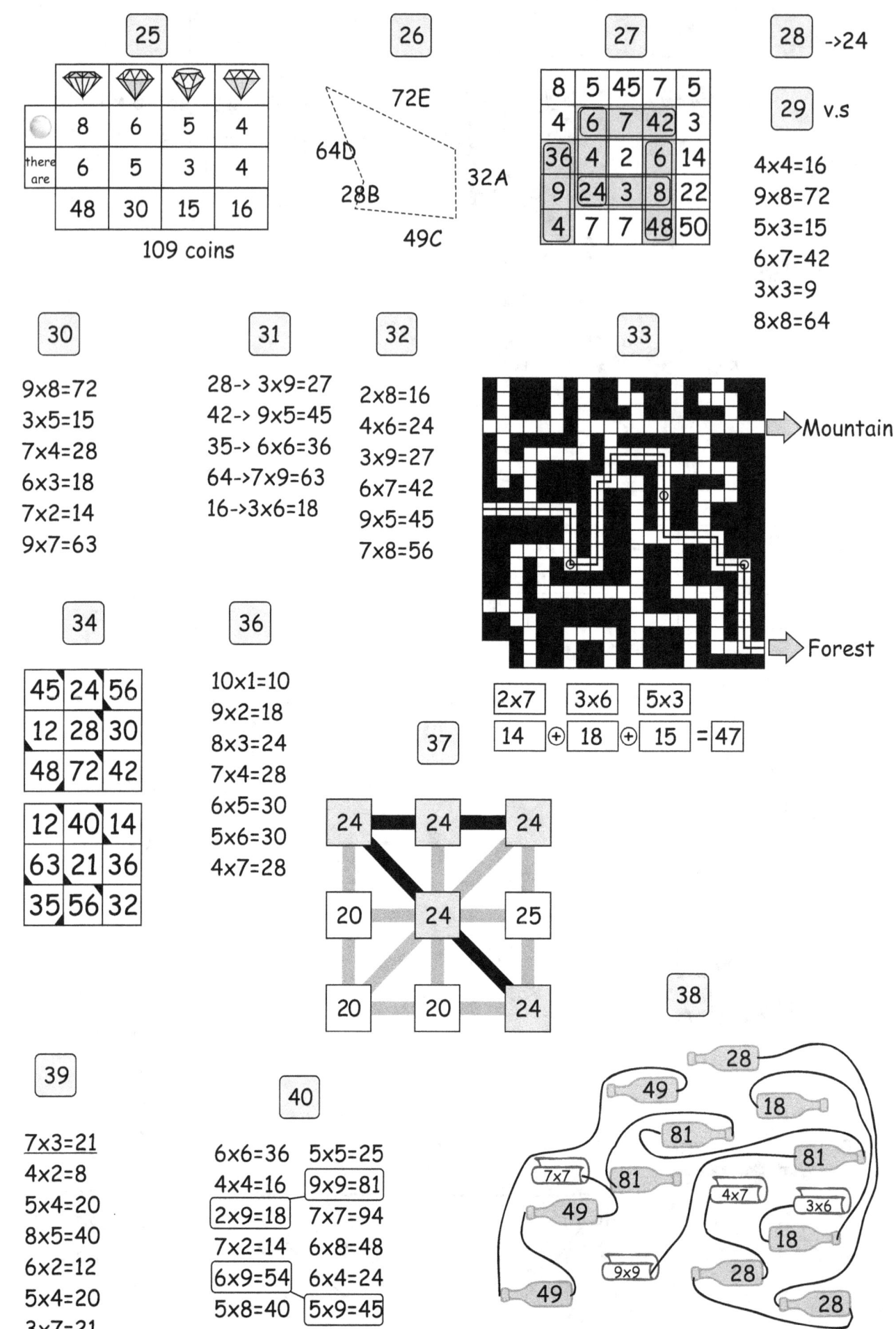

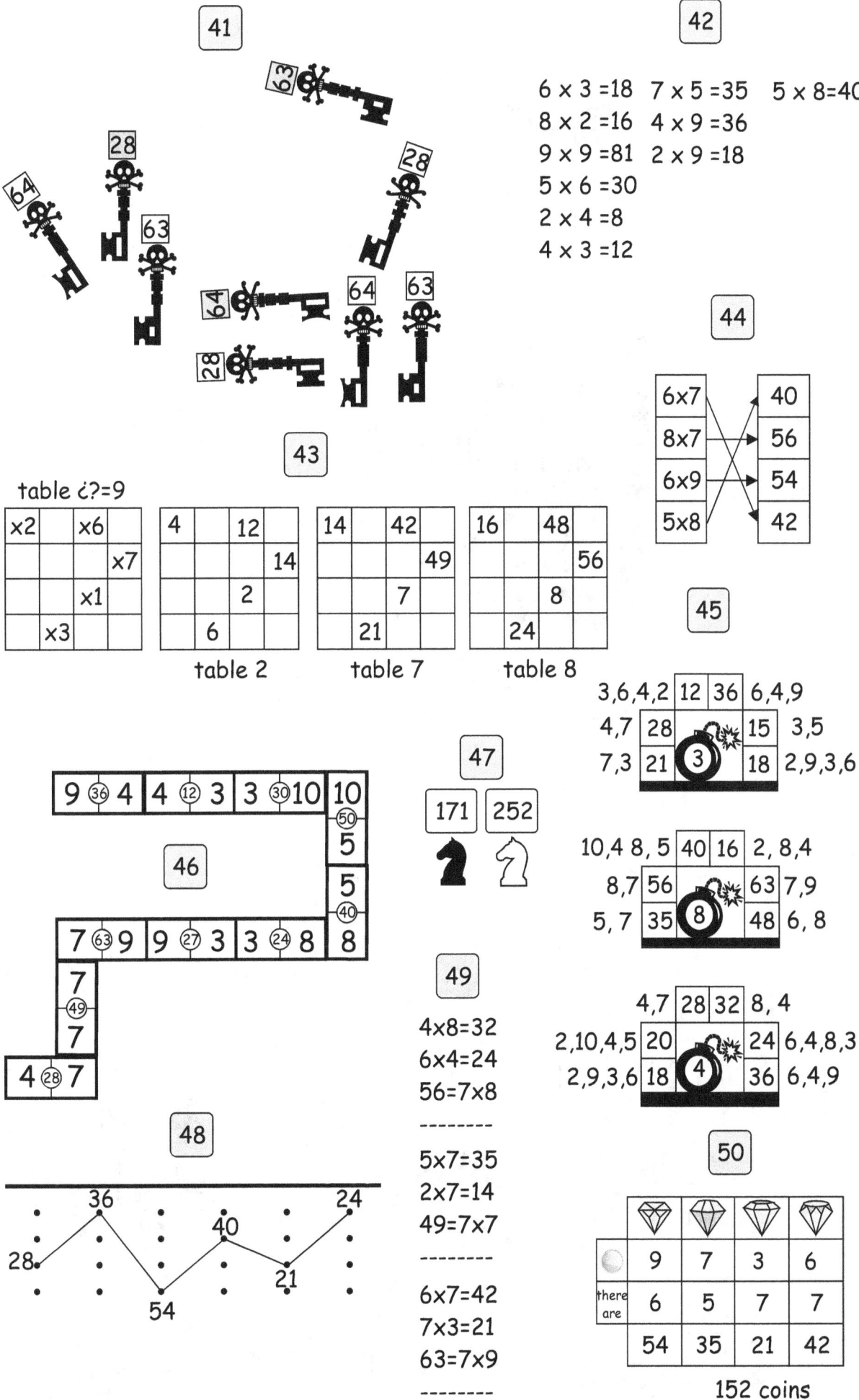

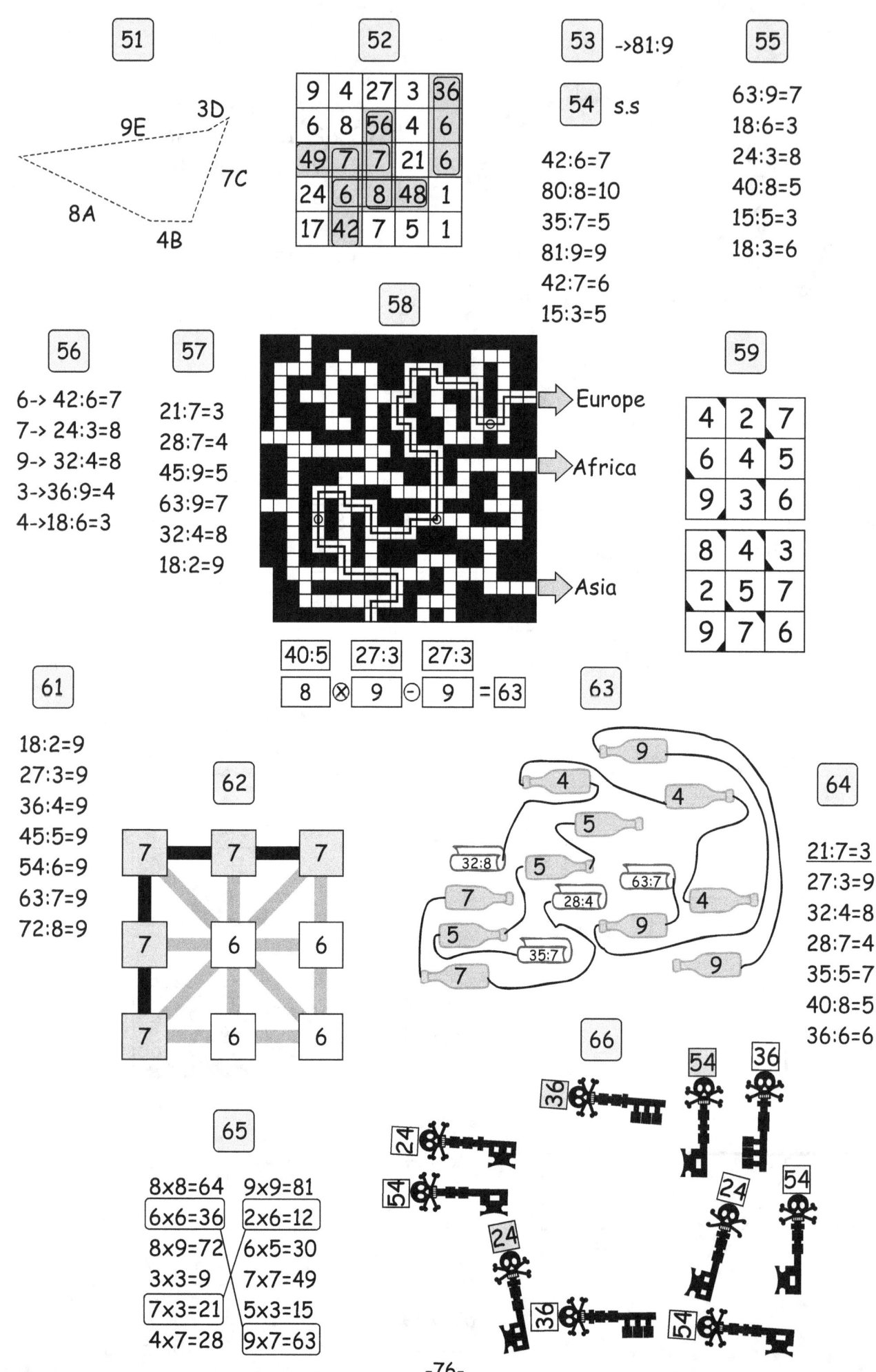

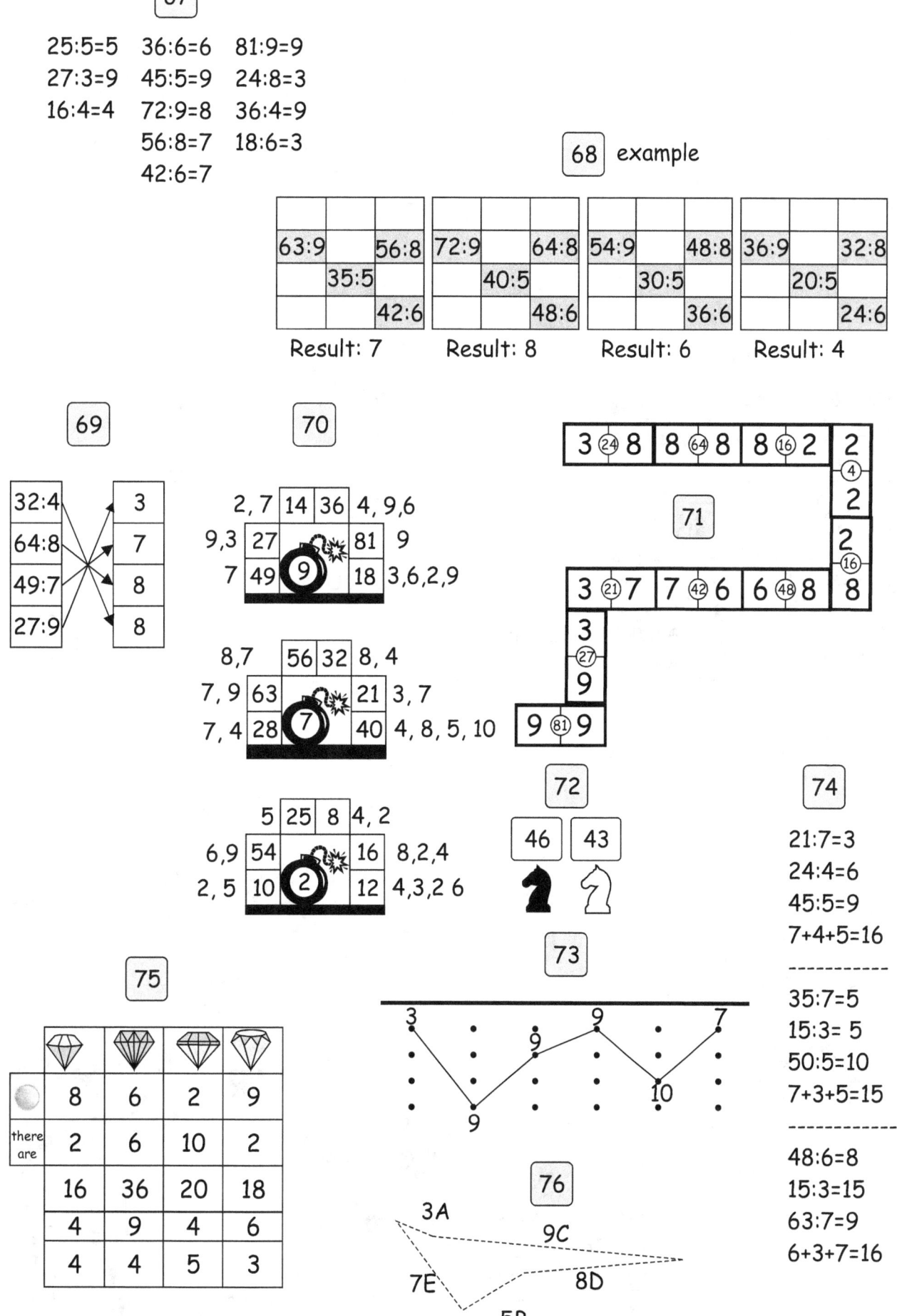

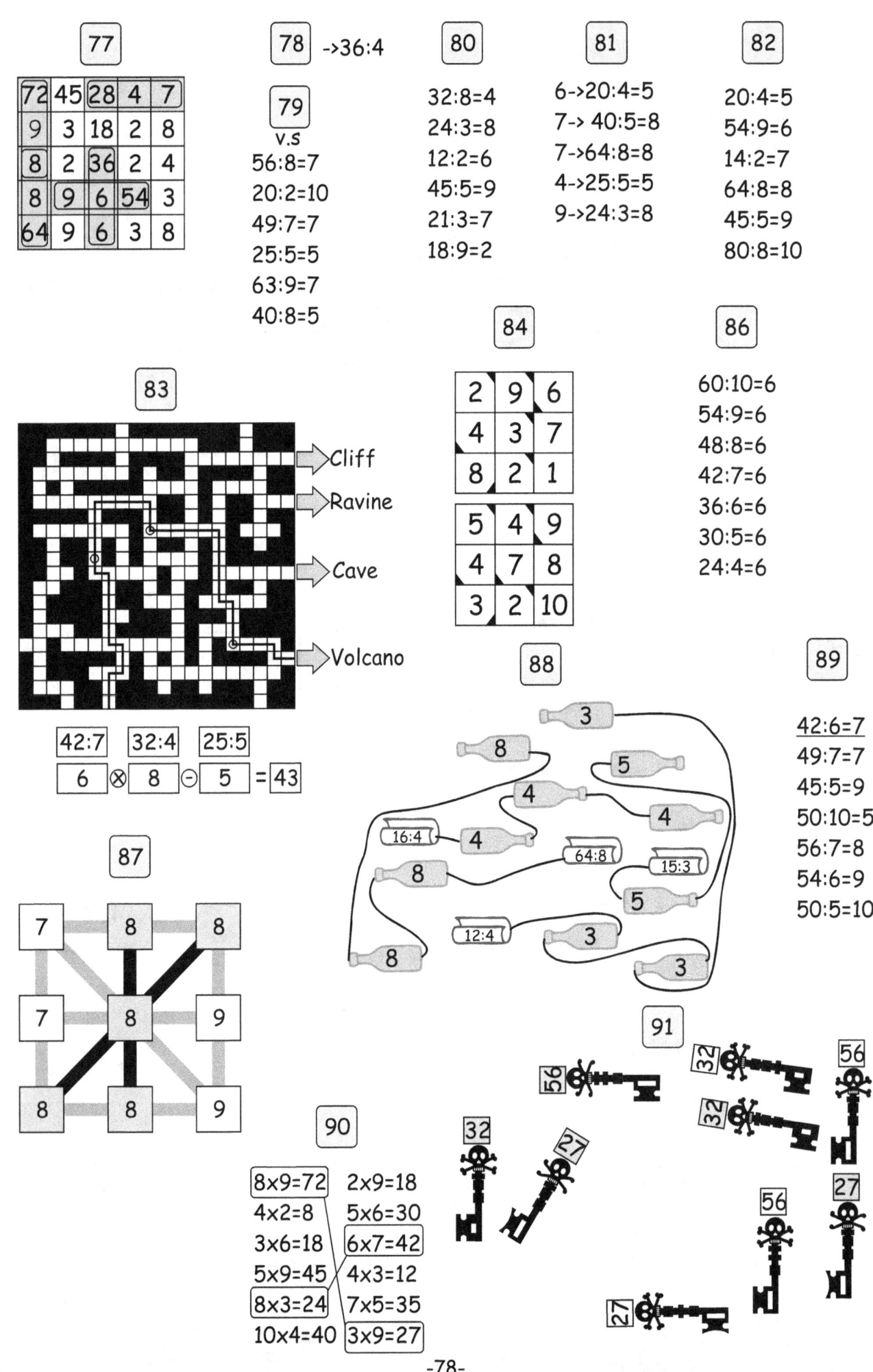

92

15:3=5 14:7=2 21:3=7
12:3=4 36:6=6 14:7=2
36:6=6 28:4=7
64:8=8
72:9=8

93 exemple

	15:5	
18:6		
		12:4
	21:7	

Result: 3

	25:5	
30:6		
		20:4
	35:7	

Result: 5

	35:5	
42:6		
		28:4
	21:3	

Result: 7

	45:5	
54:6		
		36:4
	27:3	

Result: 9

94

54:6 → 6
49:7 → 9
36:4 → 7
18:3 → 9

95

	5	25	30	
8,6	48	💣	42	6, 7
5,3	15	⑤	45	5, 9

top: 6, 5, 3, 10

	8	64	24	
8,6	48	💣	32	4, 8
2,8,4	16	⑧	40	4, 10, 8, 5

top: 6, 8, 4, 3

	10,3,6,5	30	15	3, 5
2, 10, 4, 5	20	💣	27	9, 3
3	9	③	12	2, 6, 4, 3

96

3 ⑱ 6 | 6 ㊱ 6 | 6 ㊽ 8 | 8
                                  ㊾
                                   7
                                   7
                                  ㊾
6 ㉔ 4 | 4 ⑯ 4 | 4 ㉘ 7 | 7
6
⑫
2
7 ⑭ 2

97

| 29 | 39 |

♞ ♘

98

(dot grid with connected path: 8, 2, 7, 9, 5, 6)

99

27:9=3
18:3=6
45:5=9
9+3+5=17

42:7=6
30:6=5
72:8=9
7+6+8=21

24:6=4
32:4=8
56:8=7
6+4+8=18

100

	💎	💎	💎	💎
⚪	8	3	4	6
there are	3	10	4	6
	24	30	16	36
	4	6	8	4
	6	5	2	9

www.ingramcontent.com/pod-product-compliance
Lightning Source LLC
Chambersburg PA
CBHW081454220526
45466CB00008B/2646